E 4⁰⁰

Engineering Design Process

Yousef Haik
Florida State University

THOMSON
™

BROOKS/COLE

Australia • Canada • Mexico • Singapore • Spain • United Kingdom • United States

THOMSON

BROOKS/COLE

Publisher: *Bill Stenquist*
Marketing Manager: *Tom Ziolkowski*
Marketing Assistant: *Darcie Pool*
Editorial Coordinator: *Valerie Boyajian*
Production Coordinator: *Mary Vezilich*
Print Buyer: *Vena Dyer*
Permissions Editor: *Karyn Morrison*

Production Service: *RPK Editorial*
Cover Designer: *Roy R. Neuhaus*
Cover Photos and Illustration:
Courtesy of Litespeed Titanium and Cannondale Corporation
Compositor: *Thomson Press Ltd.*
Printer: *Phoenix Color Corporation*

Printed in the United States of America
1 2 3 4 5 6 7 06 05 04 03 02

For more information about our products,
contact us at:
Thomson Learning Academic Resource Center
1-800-423-0563

For permission to use material from this text,
contact us by:
Phone: 1-800-730-2214   **Fax:** 1-800-730-2215
Web: http://www.thomsonrights.com

Library of Congress
  Control Number: 2002108119

ISBN 0-534-38014-X

Brooks/Cole – Thomson Learning
511 Forest Lodge Road
Pacific Grove, CA 93950
USA

Asia
Thomson Learning
5 Shenton Way #01-01
UIC Building
Singapore 068808

Australia
Nelson Thomson Learning
102 Dodds Street
South Melbourne,
Victoria 3205
Australia

Canada
Nelson Thomson Learning
1120 Birchmount Road
Toronto, Ontario M1K 5G4
Canada

Europe/Middle East/Africa
Thomson Learning
High Holborn House
50/51 Bedford Row
London WC1R 4LR
United Kingdom

To my family
my mentor, Marty Chen,
and my students

Brief Contents

Contents

Preface

Early human history shows that the design of artifacts was determined by human needs. For example, the knife was invented to protect humans from dangerous surroundings and Egyptians invented the wheel to facilitate travel and transportation. The truth of the dictum "necessity is the mother of invention" is demonstrated here. Design and development distinguish engineering from science. Engineers throughout history have wrestled with problems of water not being where it is needed, of minerals not being close at hand, of building materials having to be moved. Ancient engineers were often called on to devise the means for erecting great monuments, for designing defenses against enemies, and for moving people and goods across rough terrain and even rougher water. These endeavors were primarily inventive in nature and invariably approached through trial and error. However, it was not until the time of Galileo and Newton that the foundations students now learn in engineering schools were established. The Industrial Revolution brought a proliferation of new machines and manufacturing techniques and provided an impetus for the growth of science and commerce on an international scale. The history of engineering abounds with stories of success triumphing over failure.

In the early 1900s, it was common in American industry for master mechanics to invent and, subsequently, for draftsmen to copy on paper what had been synthesized experimentally in the shop. Since it was less costly and more efficient to erase rather than to remark parts in the shop, the value of synthesizing on paper was soon realized. Recent trends have been to apply theory where appropriate in the process of mechanical design. But overall, it is emphasized that all of the useful ingredients—such as various aspects of art, science, engineering, practical experience, and ingenuity—must be properly blended in the design process. A successful design is achieved when a logical procedure is followed to meet a specific need. This procedure, called the design process, is similar to the scientific method with respect to its step-by-step routine. The design process entails the following activities, all of which must be completed:

1. Identification of a need and setting of goals
2. Market analysis (search)
3. Specifications and constraints
4. Function analysis

5. Concept generation
6. Evaluation of alternatives
7. Analysis
8. Experiment
9. Marketing

Often, designs are not accomplished by an engineer simply completing the design steps in the given order. The design process holds within its structure an iterative procedure. As the engineer proceeds through the steps, new information may be discovered and new objectives may be specified, at which time the steps may require revisiting. The more time and effort an engineer spends on articulating the problem definition and understanding the needs statement, the less frequent the need for iteration.

This book is written as an introductory course in design. Students' technical capabilities are assumed to be at the level of college physics and calculus. For students with advanced technical capabilities the analysis part in the design sequence could be emphasized. Please visit the labs web page for additional information.

This book consists of three parts. Part I presents ten chapters on the design process. Chapter 1 is an overview of the design steps and serves as an introduction to the book. Chapter 2 presents a few design tools that designers must master prior to the design process. Some of these tools serve as an introduction to courses that students will encounter in future course work. Chapters 3 through 9 present the steps of the design process. The author is aware that the sequence of these steps can be changed according to instructor preference. Instructors can alter the presentation sequence without having to change the presentation material. Chapter 10 discusses issues relating to the design cost. Part II presents design projects that have been used by the author. Part III presents design labs. The purpose of these labs is to create design activities that help students, especially freshmen and sophomores, to adjust to working in teams. The first few of these labs are geared toward team building. It is anticipated that instructors may want to include other activities in their design classes. The author is building a Web page where designers and design instructors may submit their labs; the address is http://www.eng.fsu.edu/~haik/design. This web site will provide a forum for exchanging ideas on teaching design and offering relevant activities that aid students during their design carreer.

The author wishes to thank all colleagues and students who helped in producing this book, as well as all designers and design instructors who will participate in submitting design labs to the Web page. Students are encouraged to submit their comments and suggestions to the author. The author wishes to thank the following reviewers for their helpful suggestions: Thomas R. Grimm, Michigan Technological University; Jesa Kreiner, California State University—Fullerton; Thomas R. Kurfess, Georgia Tech; and Howard L. Reynolds, Texas Tech University. The author thanks Namas Chandra for his valuable contribution.

— Yousef Haik

Part I
Engineering Design Process

1 Introduction

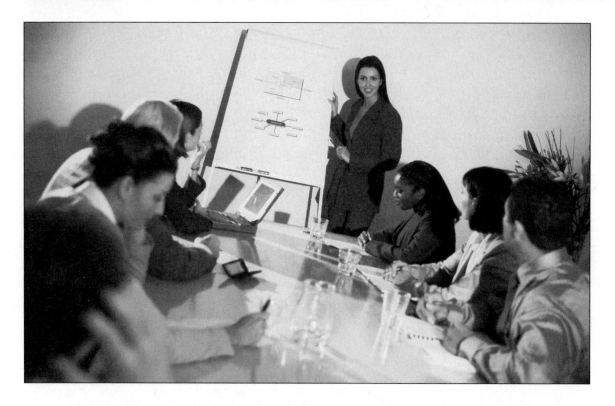

One of the first steps in the engineering design process is the design meeting. In a design meeting, engineers, technicians, and other staff members come up with solutions to fill a specific customer need. (Copyright Antonio Mo/Getty Images/PhotoDisc.)

1.1 OBJECTIVES

By the end of this chapter, you should be able to

1. Name the steps for the design process.
2. Recognize the different types of designs.
3. Define engineering design.
4. Distinguish among the different modes of design.
5. Discuss ethical problems and ethic codes.

Design is a creative process. Albert Einstein asserted that imagination is more important than knowledge, for knowledge is finite whereas imagination is infinite.

Engineering students during their training are presented with a vast amount of theoretical material and information. They only realize their weakness when they are faced with the task of logically applying what they have learned to a specific end. As long as their work is based on familiar models or previous designs, the knowledge they possess is perfectly adequate to enable them to find a solution along conventional lines. As soon as they are required to develop something already in existence to a more advanced stage, or to create something entirely new without a previous design, they will fail miserably unless they have reached a higher level of understanding. The design process was formalized to enable students to follow a systematic approach to design. The most important step of the design process is identifying the customer need. Most of the time the customer provides a generic statement of need, and it is up to the engineer to identify the real need of the customer.

For example, we are required to design a chair that can be used by a child. Clearly, all of us know how to sit on a chair, so in that perspective we know how a chair works. A chair is used for sitting. Unfortunately, this description does not say how the chair is made. What material is used? Is the chair flexible or rigid? Does the chair rotate or is it fixed? What does it mean that the chair is to be used by a child? Is safety the biggest concern? How much will the chair cost? How old is the child? and so on.

Many factors in engineering design are not based on a mathematical model, but the engineering design process is maintained to be systematic. In the previous example, we can use Newton's laws of equilibrium to describe the forces generated in the chair's legs. We can also describe the deformation of the legs when a person sits down. We can even use finite element analysis to estimate the stresses in the legs, the seat, and the back. We can also describe the make and the joints used in the chair. Some of these factors can be represented using mathematical models, but what mathematical model will describe the color of the chair, and what mathematical model will measure if the chair is safe for use by a child? The function of the chair cannot be completely presented by a mathematical model. These functions are manipulated by reasoning. Different schemes may be needed to describe a certain design (e.g., analytical models identify the geometric presentation of the artifact, economic schemes describe the cost of producing the artifact, verbal contents describe the function of the artifacts, and so on). In the designing process, mathematical modeling, although important, is not sufficient to design the artifact.

The design engineer must learn to think independently, to draw conclusions, and to combine solutions. Many believe that they can acquire this skill set by attending lectures and reading textbooks. They fail to realize that they are only accumulating one fresh item of knowledge after another. Understanding, logical deduction, and judgment cannot be conferred from outside; on the contrary, they are acquired only by diligent thinking and working with the knowledge already possessed. A basic precondition for independent design is a lively imagination. Such imagination is required to do original work.

Throughout history, humans have been successfully designing artifacts to satisfy the needs of civilization. History is full of great designs and invention. Recently, design has been driven to meet an existing requirement, reduce a hazard or an inconvenience, or develop a new approach. Not all that engineers build has become successful; occasionally, catastrophic failures occur. A few of the well-publicized disasters associated with engineering systems are as follows:

- Chernobyl nuclear power plant disaster
- Challenger space shuttle explosion
- Loss of cabin roof during flight of Boeing 737 in 1988
- Collapse of a skywalk at the Kansas City Hyatt Regency Hotel just after the hotel was opened in 1981
- Loss of an engine and subsequent crash of a DC-10 airplane in 1979

Walton lists the reason for failures in most engineering designs:

- Incorrect or overextended assumptions
- Poor understanding of the problem to be solved
- Incorrect design specifications
- Faulty manufacturing and assembly
- Error in design calculations
- Incomplete experimentation and inadequate data collection
- Errors in drawings
- Faulty reasoning from good assumptions

1.2 DEFINTION OF ENGINEERING DESIGN

Dym defines design as the systematic, intelligent generation and evaluation of specifications for artifacts whose form and function achieve stated objectives and satisfy specified constraints. A formal definition of engineering design is found in the curriculum guidelines of the Accreditation Board for Engineering and Technology (ABET). The ABET definition states that engineering design is the process of devising a system, component, or process to meet desired needs. It is a decision-making process (often iterative), in which the basic sciences, mathematics, and engineering sciences are applied to convert resources optimally to meet a stated objective. Among the fundamental elements of the design process is the establishment of objectives and criteria, synthesis, analysis, construction, testing, and evaluation. The engineering design component of a curriculum must include most of the following features: development of student creativity, use of open-ended problems, development and use of modern design theory and methodology, formulation of design problem statement and specifications, production processes, concurrent engineering design, and detailed system description. Further, it is essential to include a variety of realistic constraints such as economic factors, safety, reliability, aesthetics, ethics, and social impact.

1.3   DESIGN LEVELS

As in any field of human activity, there are different degrees of difficulty. In design these stages are adaptive design, developed design, and new design.

- *Adaptive design:* In the great majority of instances, the designer's work will be concerned with adaptation of existing designs. There are branches of manufacturing in which development has practically ceased, so that there is hardly anything left for the designer to do except make minor modifications, usually in the dimensions of the product. Design activity of this kind demands no special knowledge or skill, and the problems presented are easily solved by a designer with ordinary technical training.
- *Development design:* Considerably more scientific training and design ability are needed for development design. The designer starts from an existing design, but the final outcome may differ markedly from the initial product.
- *New design:* Only a small number of designs are new designs.

1.4   DESIGN PROCESS

To design is to create new product that turns into profit and benefits society in some way. The ability to design requires both science and art. The science can be learned through a systematic process (outlined in this chapter), experience, and problem-solving technique (all of which will be mastered during your college education). The art is gained by practice and a total dedication to becoming proficient. The design of a device or system can be done in one of two ways:

1. *Evolutionary change:* A product is allowed to evolve over a period of time with only slight improvement. This is done when there is no competition. The creative capabilities of the designer are limited.
2. *Innovation:* Rapid scientific growth and technological discoveries as well as competition among companies for their slice of the market have placed a great deal of emphasis on new products, which draw heavily on innovation. The creative skills and analytical ability of the design engineer play an important role.

Proficient designers control evolution and innovation so they occur simultaneously. Although the emphasis is on innovation, designers must test their ideas against prior design. Engineers can design for the future but must base results on the past.

This section provides a brief description of the design process. The design process is a sequence of events that help define the stages of design and unfold the

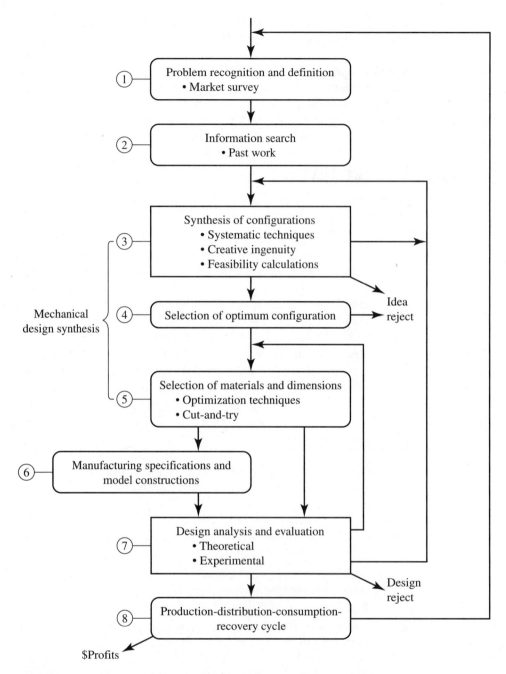

Figure 1.1   Design process map. (From Johnson, 1978.)

design in a systematic manner. Various authors have offered charts that describe the design process. Figures 1.1 through 1.3 show some of these attempts.

The difference in these charts is in sequence names. Careful examination of the charts leads to identification of the same stages. Some stages are combined in one

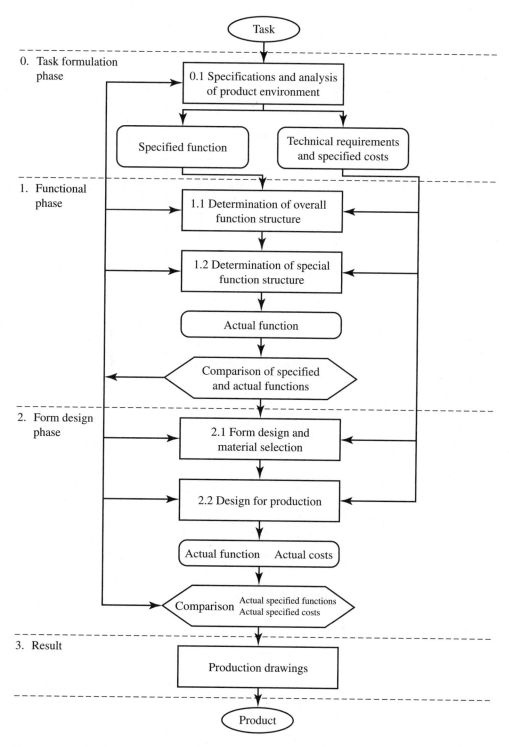

Figure 1.2 Design process map. (From Dym 1994.)

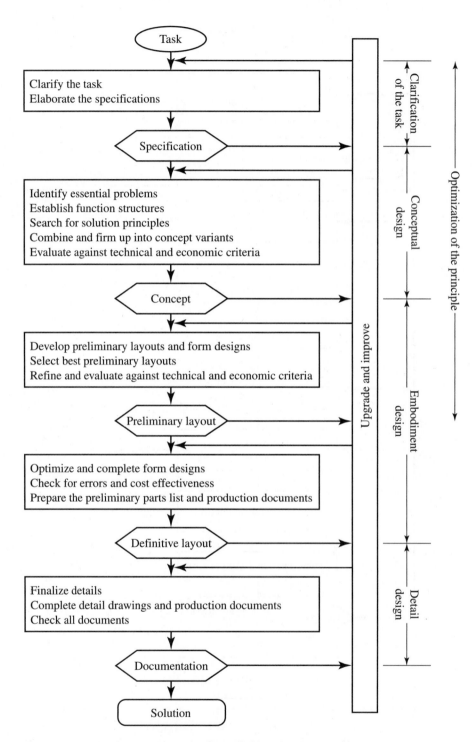

Figure 1.3   Design process map. (From Pahl and Beitz 1996.)

process. In Figure 1.4 the steps are iterative and require a series of decisions to move the design along. More often a design oscillates back and forth between stages until it reaches an acceptable form and can move to the next stage. A brief description of these stages is provided in the following subsections.

1.4.1 Identifying Customer Need

The need for a new design can be generated from several sources, including the following:

- *Client request:* In a design company a client may submit a request for developing an artifact. It is often unlikely that the need will be expressed clearly. The client may know only the type of product that he or she wants; for example, "I need a safe ladder."

- *Modification of an existing design:* Often a client asks for a modification of an existing artifact to make it simpler and easy to use. In addition, companies may want to provide customers with new, easy-to-use products. For example, in a market search you may notice many brand names for coffee makers and the differences among them, such as shape, material used, cost, or special features. As another example, Figures 1.5 through 1.8 demonstrate design developments for paperclips. Each of these new designs has its own advantages over the other clips. For example, the end-

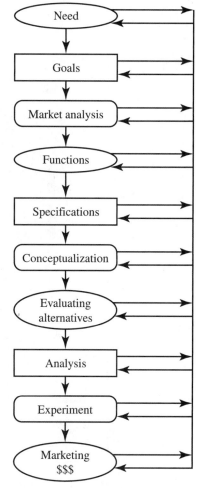

Figure 1.4 Design process.

less filament paperclip can be used from either side of the clip. One may argue that the different designs are based on the human evolution of designs and birth of new ideas; however, the major driving force for the renovation of designs is to keep companies in business. The first patent for a paperclip was filed in 1899 and the latest was filed in 1994—a hundred years of paperclip development and innovations.

- *Generation of new product:* In all profit-oriented industries the attention, talent, and abilities of management, engineering, production, inspection, advertising, marketing, sales, and servicing are focused on causing the product to return profit for the company and, in turn, for company stockholders. Unfortunately, sooner or later every product is preempted by another or degenerates into profitless price competition. For an industry to survive in today's world, it must continue to grow; it cannot afford to remain static. This growth, throughout history, has been built on new products. New products

Figure 1.5 William Middelbrook's patent for a machine for making paperclips (1899).

have a characteristic life cycle pattern in sales volume and profit margins, as shown in Figure 1.9. A product will peak out when it has saturated the market and then begin to decline. It is obvious that an industry must seek out and promote a flow of new product ideas. These new products are usually protected by applying for patents.

1.4.2 Defining Goals

In this process, the designer defines what must be done to resolve the need. The definition is a general statement of the desired end product. Many of the difficulties encountered in design may be traced to poorly stated goals, or goals that were hastily written and resulted in confusion or too much flexibility.

In almost all cases the client request comes in a vague verbal statement such as, "I need an aluminum can crusher." or "I need a safe ladder." Designers must

Figure 1.6 Patent for Gothic-style paperclip, issued in 1934.

recognize that customer needs are not the same as product specifications. Needs should be expressed in functional terms. Customers will offer solutions; designers must determine the real needs and act accordingly. During the customer interview, listen carefully to what the customer has to say. The designer's function is to clarify the client's design requirements. An objective tree may be constructed for clarification. Often the need statement and goals are combined into one process. An objective tree is a tool used by designers to organize the customer's wants into categories; refer to Chapter 3 for details.

1.4.3 Market Analysis

Designers must locate what is already available in the market and what they have to offer. Information gathering is a vital task. Some companies hire design engineers so they can get away from paying royalties to patent holders. Design engineers may consult the following sources to determine market availability:

Figure 1.7 Patent for an endless filament paperclip, issued to Charles T. Link (1991).

- Technical and trade journals
- Abstracts
- Research reports
- Technical libraries
- Catalog of component suppliers
- U.S. Patent Office
- The Internet

The information gathered may reveal an available design solution and the hardware to accomplish the goal. Sometimes the goal may be altered to produce a requested product or abandoned if the product already exists. Knowledge of existing products will save the designer and client time and money. Once the designer determines what is in the market, creativity should be directed toward generating alternatives.

United States Patent [19]

Froehlich et al.

[11] Patent Number: 5,329,672
[45] Date of Patent: Jul. 19, 1994

[54] METAL WIRE PAPER CLIP STRUCTURE

[76] Inventors: Linda A. Froehlich; Richard D. Froehlich, both of 3151 Keiners La., Pittsburgh, Pa. 15205

[21] Appl. No.: 604,970

[22] Filed: Oct. 29, 1990

Related U.S. Application Data

[63] Continuation of Ser. No. 45,452, May 4, 1987, abandoned, which is a continuation of Ser. No. 764,366, Aug. 12, 1985, abandoned.

[51] Int. Cl.5 .... B42F 1/04
[52] U.S. Cl. 24/67.9
[54] Field of Search 24/67.9, 546, 545, 547

[56] References Cited

U.S. PATENT DOCUMENTS

242,892 11/1903 McGill 24/547
342,893 11/1903 McGill .... 24/547 X
2,399,649 1/1943 Comley 24/547
4,017,377 4/1977 Winer et al. .... 24/547 X
4,597,139 7/1986 Lee 24/546

FOREIGN PATENT DOCUMENTS

445143 5/1936 Fed. Rep. of Germany 24/544

Primary Examiner—James R. Brittain
Attorney, Agent, or Firm—Raymond N. Baker

[57] ABSTRACT

Improved paper clip is formed from a single piece of spring-quality metal wire to have an elongated U-shaped inner loop nested within an elongated U-shaped outer loop. Each loop includes a free leg and a connector leg; the longitudinally-extending leg portions of the free leg and connector leg of each respective loop are of substantially the same length. The U-shaped loops are joined together by an arcuately-curved interconnector extending between the connector leg of each loop. Distal ends of the free leg of each loop are located contiguous to such arcuately-curved interconnector which defines one longitudinal end of the paper clip with such new configurational and other features. damage to outer surfaces of stacked paper by such distal ends is avoided and clasping force is applied along substantially the full length of such free legs during use of the improved clip.

5 Claims, 2 Drawing Sheets

Figure 1.8 Patent for a large paperclip made of spring wire, issued to Linda and Richard Froehlich (1994).

1.4.4 Establishing Functions

Recognizing the generality of the need statement and where the problem/need stands in the whole system is a fundamental element in the design process. There is a big difference between being asked to design a car suspension system and designing a car. It is useful to consider the level at which the designer is asked to work. It is also useful to identify the functions instead of the potential solutions. For example, a blood bank approached a designer to find a solution for its frequently broken centrifuge. When the centrifuge breaks down, the blood separation unit shuts down. The blood bank has tried replacing the centrifuge, but after a few months it breaks down. The need statement is to fix the centrifuge in such a way that it will reduce the break-down time. A designer who recognizes the design process will not jump on fixing the centrifuge; he will ask the following question: "What is the function we try to accomplish with the centrifuge?" The answer is to separate blood by enhancing the gravitational pull. Then the function is to separate blood cells from

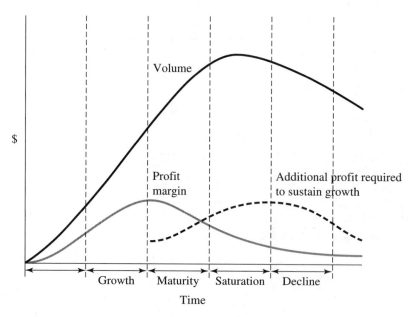

Figure 1.9   Basic life cycle of new product.

the whole blood. This clear statement enables the designer to find alternative solutions other than the centrifuge. Often the functions will be divided into subfunctions, and they will define the requirements of the artifact.

In another example, if you were required to design a lawn mower, the function should be defined as a method for shortening the grass. The difference between two definitions of a lawn mower (designing a lawn mower or method to shorten grass) is that one may limit your imagination and creativity while the other may give you free reign for creativity.

1.4.5   Task Specifications

This stage requires the designer to list all pertinent data and parameters that tend to control the design and guide it toward the desired goal; it also sets limits on the acceptable solutions. It should not be defined too narrowly because the designer will eliminate acceptable solutions; however, it cannot be too broad or vague because this will leave the designer with no direction to satisfy the design goal. Let's reconsider the lawn mower example and set some task specifications.

1. It should be safe to operate, especially when used near children.
2. It should be easy to operate by an average person.
3. Types of power to be considered:
   a. Battery
   b. Electricity
   c. 2-cycle or 4-cycle internal combustion engine
   d. Manual

4. The device should be easily stored in a garage.
5. The device should allow combinations of lawn tasks:
   a. Collecting leaves and mowing the lawn
   b. Chopping the leaves and fertilizing
   c. Chopping branches for mulch
6. The device should use available hardware and require sheet metal, fiber glass, or plastic.
7. Material should be selected based on cost, manufacturability, strength, appearance, and ability to withstand varying weather conditions.
8. The device must be reliable and not require frequent maintenance.
9. The selling price should be lower than that in the market.

1.4.6   Conceptualization

The process of generating alternative solutions to the stated goal in the form of concepts requires creative ability. The conceptualization starts with generating new ideas. In this stage the designer must review the market analysis and the task specifications as he or she engages in the process of innovation and creativity. This activity usually requires free-hand sketches for producing a series of alternative solutions. The alternatives do not need to be worked out in detail but are recorded as possibilities to be tested. Alternatives to perform the functions should be listed in an organized fashion. For example, to design a novel transportation system, a designer may list the methods as follows:

1. Natural way
   a. Human
      i. Walk
      ii. Swim
   b. Animal
      i. Ride
      ii. Pulled in a cart
2. With aids
   a. Land
      i. Bike
      ii. Skate
   b. Water
      i. Canoe
      ii. Tube
   c. Air
      i. Kite
   d. Mechanical
      i. Land
         • Car
         • Train
         • Tube

    ii. Water
- Ship
- Sled

    iii. Air
- Plane
- Rocket

1.4.7 Evaluating Alternatives

Once a number of concepts have been generated in sufficient detail, a decision must be made about which one or ones will enter the next, most expensive, stages of the design process. An excellent technique to guide the designer in making the best decision regarding these alternatives is a scoring matrix, which forces a more penetrating study of each alternative against a specified criteria.

1.4.8 Analysis and Optimization

Once a concept has been chosen as a possible solution for the stated goal, it must be tested against the physical laws. The manufacturability of the chosen product also must be checked to ensure its usefulness. A product may satisfy the physical laws, but if it cannot be manufactured it is a useless product. This stage is put in iterative sequencing with the conceptualization. Often analysis requires a concept to be altered or redefined, then reanalyzed so that the design is constantly shifted between analysis and concept. Analysis starts with estimation and is followed by order of magnitude calculation. Estimation is an educated guess based on experience. Order of magnitude analysis is a rough calculation of the specified problem. The order of magnitude does not provide an exact solution, but it gives the order in which the solution should be expected.

1.4.9 Experiment

The experiment stage in engineering design requires that a piece of hardware is constructed and tested to verify the concept and analysis of the design as to its work ability, durability, and performance characteristics. Here the design on paper is transformed into a physical reality. Three techniques of construction are available to the designer:

1. *Mock-up:* The mock-up is generally constructed to scale from plastics, wood, cardboard, and so forth. The mock-up is often used to check clearance, assembly technique, manufacturing considerations, and appearance. It is the least expensive technique, provides the least amount of information, and is quick and relatively easy to build.
2. *Model:* This is a representation of the physical system through a mathematical similitude. Four types of models are used to predict behavior of the real system:
   a. A true model is an exact geometric reproduction of the real system, built to scale and satisfying all restrictions imposed in the design parameters.
   b. An adequate model is so constructed to test specific characteristics of the design.

c. A distorted model purposely violates one or more design conditions. This violation is often required when it is difficult to satisfy the specified conditions.

d. Dissimilar models bear no apparent resemblance to the real system, but through appropriate analogies give accurate information on behavioral characteristics.

3. *Prototype:* This is the most expensive experimental technique and the one producing the greatest amount of useful information. The prototype is the constructed, full-scale working physical system. Here the designer sees his or her idea come to life and learns about such things as appropriate construction techniques, assembly procedures, work ability, durability, and performance under actual environmental conditions.

As a general rule, when entering the experimental stage of the design process, one should first deal with the mock-up, then the model, and finally the prototype (after the mock-up and model have proven the real worth of the design), to allow beneficial interaction with concept and analysis.

1.4.10　Marketing

This stage requires specific information that defines the device, system, or process. Here the designer is required to put his or her thoughts regarding the design on paper for the purpose of communication with others. Communication is involved in selling the idea to management or the client, directing the shop on how to construct the design, and serving management in the initial stages of commercialization. The description should take the form of one of the following:

1. A report containing the following: a detailed description of the device, how it satisfies the need and how it works, a detailed assembly drawing, specifications for construction, a list of standard parts, a cost breakdown, and any other information that will ensure that the design will be understood and constructed exactly as the designer intended.

2. A flyer containing a list of the special features that the design can provide, advertisements, promotional literature, market testing, and so forth.

1.5　PROFESSIONALISM AND ETHICS

Professionalism is a way of life. A professional person is one who engages in an activity that requires a specialized and comprehensive education and is motivated by a strong desire to serve humanity. The work of engineers generally affects the day-to-day life of all humans. Developing a professional frame of mind begins with your engineering education. Several authors have attempted to provide general rules of ethics that can be followed in all circumstances. However, they have failed, as some of the rules will be problem-, profession-, and situation-dependent. Other authors have discussed the ethics as a set of moral values that are associated with culture and religion. Several engineering societies have developed a code of ethics that must be followed by engineers.

Ethics is a complex and emerging subject. Because of its importance, the author has designed a lab that deals with ethics (see p. 232). Students should do that lab. For additional reading, several online resources deal with ethics, including the following:

- http://web.mit.edu/ethics/www/essays/probcase.html
- http://www.onlineethics.org

1.5.1 ASME Code of Ethics

The American Society of Mechanical Engineers (ASME) requires ethical practice by each of its members and has adopted the following Code of Ethics of Engineers as referenced in the ASME Constitution, Article C2.1.1.

Engineers uphold and advance the integrity, honor and dignity of the engineering profession by:

I. Using their knowledge and skill for the enhancement of human welfare;

II. Being honest and impartial, and serving with fidelity the public, their employers and clients; and

III. Striving to increase the competence and prestige of the engineering profession.

The Fundamental Canons

1. Engineers shall hold paramount the safety, health and welfare of the public in the performance of their professional duties.

2. Engineers shall perform services only in the areas of their competence.

3. Engineers shall continue their professional development throughout their careers and shall provide opportunities for the professional and ethical development of those engineers under their supervision.

4. Engineers shall act in professional matters for each employer or client as faithful agents or trustees, and shall avoid conflicts of interest or the appearance of conflicts of interest.

5. Engineers shall build their professional reputation on the merit of their services and shall not compete unfairly with others.

6. Engineers shall associate only with reputable persons or organizations.

7. Engineers shall issue public statements only in an objective and truthful manner.

The ASME criteria for interpretation of the Canons are guidelines and represent the objectives toward which members of the engineering profession should strive. They are principles, which an engineer can reference, in specific situations. In addition, they provide interpretive guidance to the ASME Board on Professional Practice and Ethics on the Code of Ethics of Engineers.

1. Engineers shall hold paramount the safety, health and welfare of the public in the performance of their professional duties.
   (a) Engineers shall recognize that the lives, safety, health and welfare of the general public are dependent upon engineering judgments, decisions and practices incorporated into structures, machines, products, processes and devices.

(b) Engineers shall not approve or seal plans and/or specifications that are not of a design safe to the public health and welfare and in conformity with accepted engineering standards.

(c) Whenever the engineers' professional judgments are over ruled under circumstances where the safety, health, and welfare of the public are endangered, the engineers shall inform their clients and/or employers of the possible consequences.

  i. Engineers shall endeavor to provide data such as published standards, test codes, and quality control procedures that will enable the users to understand safe use during life expectancy associated with the designs, products, or systems for which they are responsible.

  ii. Engineers shall conduct reviews of the safety and reliability of the designs, products, or systems for which they are responsible before giving their approval to the plans for the design.

  iii. Whenever engineers observe conditions, directly related to their employment, which they believe will endanger public safety or health, they shall inform the proper authority of the situation.

(d) If engineers have knowledge of or reason to believe that another person or firm may be in violation of any of the provisions of these Canons, they shall present such information to the proper authority in writing and shall cooperate with the proper authority in furnishing such further information or assistance as may be required.

2. Engineers shall perform services only in areas of their competence.

(a) Engineers shall undertake to perform engineering assignments only when qualified by education and/or experience in the specific technical field of engineering involved.

(b) Engineers may accept an assignment requiring education and/or experience outside of their own fields of competence, but their services shall be restricted to other phases of the project in which they are qualified. All other phases of such project shall be performed by qualified associates, consultants, or employees.

3. Engineers shall continue their professional development throughout their careers, and should provide opportunities for the professional and ethical development of those engineers under their supervision.

4. Engineers shall act in professional matters for each employer or client as faithful agents or trustees, and shall avoid conflicts of interest or the appearance of conflicts of interest.

(a) Engineers shall avoid all known conflicts of interest with their employers or clients and shall promptly inform their employers or clients of any business association, interests, or circumstances which could influence their judgment or the quality of their services.

(b) Engineers shall not undertake any assignments which would knowingly create a potential conflict of interest between themselves and their clients or their employers.

(c) Engineers shall not accept compensation, financial or otherwise, from more than one party for services on the same project, or for services pertaining to the same project, unless the circumstances are fully disclosed to, and agreed to, by all interested parties.

(d) Engineers shall not solicit or accept financial or other valuable considerations, for specifying products or material or equipment suppliers, without disclosure to their clients or employers.

(e) Engineers shall not solicit or accept gratuities, directly or indirectly, from contractors, their agents, or other parties dealing with their clients or employers in connection with work for which they are responsible. Where official public policy or employers' policies tolerate acceptance of modest gratuities or gifts, engineers shall avoid a conflict of interest by complying with appropriate policies and shall avoid the appearance of a conflict of interest.

(f) When in public service as members, advisors, or employees of a governmental body or department, engineers shall not participate in considerations or actions with respect to services provided by them or their organization(s) in private or product engineering practice.

(g) Engineers shall not solicit an engineering contract from a governmental body or other entity on which a principal, officer, or employee of their organization serves as a member without disclosing their relationship and removing themselves from any activity of the body which concerns their organization.

(h) Engineers working on codes, standards or governmental sanctioned rules and specifications shall exercise careful judgment in their determinations to ensure a balanced viewpoint, and avoid a conflict of interest.

(i) When, as a result of their studies, engineers believe a project(s) will not be successful, they shall so advise their employer or client.

(j) Engineers shall treat information coming to them in the course of their assignments as confidential, and shall not use such information as a means of making personal profit, if such action is adverse to the interests of their clients, their employers or the public.

   i. They will not disclose confidential information concerning the business affairs or technical processes of any present or former employer or client or bidder under evaluation, without his consent, unless required by law or court order.

   ii. They shall not reveal confidential information or finding of any commission or board of which they are members unless required by law or court order.

   iii. Designs supplied to engineers by clients shall not be duplicated by the engineers for others without the express permission of the client(s).

(k) Engineers shall act with fairness and justice to all parties when administering a construction (or other) contract.

(l) Before undertaking work for others in which engineers may make improvements, plans, designs, inventions, or other records which may justify seeking copyrights, patents, or proprietary rights, engineers shall enter into positive agreements regarding the rights of respective parties.

(m) Engineers shall admit their own errors when proven wrong and refrain from distorting or altering the facts to justify their mistakes or decisions.

(n) Engineers shall not accept professional employment or assignments outside of their regular work without the knowledge of their employers.

(o) Engineers shall not attempt to attract an employee from other employers or from the market place with false or misleading representations.

5. Engineers shall build their professional reputation on the merit of their services and shall not compete unfairly with others.

(a) Engineers shall negotiate contracts for professional services on the basis of demonstrated competence and qualifications for the type of professional service required.

(b) Engineers shall not request, propose, or accept professional commissions on a contingent basis if, under the circumstances, their professional judgments may be compromised.

(c) Engineers shall not falsify or permit misrepresentation of their, or their associates', academic or professional qualifications. They shall not misrepresent or exaggerate their degrees of responsibility in or for the subject matter of prior assignments. Brochures or other presentations used to solicit personal employment shall not misrepresent pertinent facts concerning employers, employees, associates, joint ventures, or their accomplishments.

(d) Engineers shall prepare articles for the lay or technical press which are only factual. Technical Communications for publication (theses, articles, papers, reports, etc.) which are based on research involving more than one individual (including students and supervising faculty, industrial supervisor/researcher or other co-workers) must recognize all significant contributors. Plagiarism, the act of substantially using another's ideas or written materials without due credit, is unethical.

(e) Engineers shall not maliciously or falsely, directly or indirectly, injure the professional reputation, prospects, practice or employment of another engineer, nor shall they indiscriminately criticize another's work.

(f) Engineers shall not use the equipment, supplies, laboratory or office facilities of their employers to carry on outside private practice without consent.

6. Engineers shall associate only with reputable persons or organizations.

(a) Engineers shall not knowingly associate with or permit the use of their names or firm names in business ventures by any person or firm which they know, or have reason to believe, is engaging in business or professional practices of a fraudulent or dishonest nature.

(b) Engineers shall not use association with non-engineers, corporations, or partnerships to disguise unethical acts.

7. Engineers shall issue public statements only in an objective and truthful manner.

(a) Engineers shall endeavor to extend public knowledge, and to prevent misunderstandings of the achievements of engineering.

(b) Engineers shall be completely objective and truthful in all professional reports, statements or testimony. They shall include all relevant and pertinent information in such reports, statements or testimony.

(c) Engineers, when serving as expert or technical witnesses before any court, commission, or other tribunal, shall express an engineering opinion only when it is founded on their adequate knowledge of the facts in issue, their background of technical competence in the subject matter, and their belief in the accuracy and propriety of their testimony.

(d) Engineers shall issue no statements, criticisms, or arguments on engineering matters which are inspired or paid for by an interested party, or parties; unless they preface their comments by identifying themselves, by disclosing the identities of the party or parties on whose behalf they are speaking, and by revealing the existence of any financial interest they may have in matters under discussion.

(e) Engineers shall be truthful in explaining their work and merit, and shall avoid any act tending to promote their own interest at the expense of the integrity and honor of the profession or another individual.

8. Engineers accepting membership in The American Society of Mechanical Engineers by this action agree to abide by this Society Policy on Ethics and procedures for its implementation.

1.6 PROBLEMS

1.6.1 Team Activities

1. List the steps in the design process.
2. Define each step in the design process.
3. What is the difference between customer statement and problem definition?
4. What is the difference between the specification step and defining the problem step in the design process?
5. What is function analysis and how is it different from the definition of the problem?
6. List three factors that market analysis achieves.
7. Why does function analysis precede the conceptualization step?

8. Give examples of a mock-up.
9. Why is scheduling important?

1.6.2 Individual Activities

1. Figure 1.10 shows the percent of cost committed and incurred as a function of time during the design of a product. The committed cost is the amount of money allocated for the manufacturing of the product, while the incurred cost is the amount of money spent on the design.
   (a) Explain your findings from the figure.
   (b) Discuss why it is more important to spend time and money on the early stages of design than on the late stages of design.

Manufacturing cost during design

Figure 1.10 Manufacturing cost.

2. Describe in detail the various types of design.
3. List several common sources of engineering failures.
4. Explain the difference between engineering design and the engineering design process.
5. Identify several commonalities and differences among Figures 1.1 through 1.4.
6. List four factors that may be used to determine quality, and discuss the following statement in light of your listing: "Quality cannot be built into a product unless it is designed into it."
7. Consider the following two statements:
   • What size SAE grade 5 bolt should be used to fasten together two pieces of 1045 sheet steel, each 4 mm thick and 6 cm wide, which are lapped over each other and loaded with 100 N?

- Design a joint to fasten together two pieces of 1045 sheet steel, each 4 mm thick and 6 cm wide, which are lapped over each other and loaded with 100 N.
  - (a) Explain the difference between the two statements.
  - (b) Change a problem from one of your engineering science/physics classes into a design problem.

8. Explain the engineering design process to a high school student.
9. Define professionalism.
10. Read through the ASME code of ethics and list three findings.
11. The following ethical scenario was obtained from http://www.cwru.edu/wwwethics

   You are an engineer charged with performing safety testing and obtaining appropriate regulatory agency or outside testing laboratory approvals of your company's product. The Gee Whiz Mark2 (GWM2) has been tested and found compliant with both voluntary and mandatory safety standards in North America and Europe. Because of a purchase-order error and subsequent oversights in manufacture, 25,000 units of GWM2 were built that are not compliant with any of the North American or European safety standards. A user would be more vulnerable to electric shock than from a compliant unit. Under some plausible combinations of events the bad unit user could be electrocuted. Retrofitting these products to make them compliant is not feasible because the rework costs would exceed the profit margin by far. Comaney agrees that because of this defect the agency safety labels will not be attached to the bad units, as per the requirements of the several agencies. Only two options exist: (a) Scrap the units and take the loss, or (b) sell the units. An employee of the company notes that many countries have no safety standards of any kind for this type of product. It is suggested that the bad units be marketed in these countries. It is pointed out that many of these nations have no electrical wiring code; or if codes exist they are not enforced. The argument is thus advanced that the bad GWM2 units are no worse than the modus operandi of the electrical practice of these countries and their cultural values. Assume that no treaties or export regulations would be violated.
   - (a) What would be your recommendation?
   - (b) Suppose one of the countries under consideration was the country of origin for you or your recent ancestors. Would this affect your recommendations?
   - (c) Now suppose you are not asked for a recommendation, only an opinion. What is your response?
   - (d) Suppose it is suggested that the bad units be sold to a third party, who would very likely sell the units to these countries. Your comments?
   - (e) You are offered gratis one of the bad units for your use at home, provided that you sign a release indicating your awareness of the condition of the unit and that it is given to you as a test unit. Assume you cannot retrofit it and that the product could be very useful to you. Would you accept the offer?
   - (f) Suppose it is suggested that the offer stated in part (e) be made to all employees of the company. Your comments?

1.7 SELECTED BIBLIOGRAPHY

1. Ambrose, S. A., and Amon, C. H. "Systematic Design of a First-Year Mechanical Engineering Course at Carnegie Mellon University," *Journal of Engineering Education*, pp. 173–181, 1997.
2. Bucciarelli, L. L. *Designing Engineers*. Cambridge, MA: MIT Press, 1996.
3. Burger, C. P. "Excellence in Product Development through Innovative Engineering Design," *Engineering Productivity and Valve Technology*, pp. 1–4, 1995.
4. Burghardt, M. D. *Introduction to Engineering Design and Problem Solving*. New York: McGraw-Hill, 1999.
5. Cross, N. *Engineering Design Methods: Strategies for Product Design*. New York: Wiley, 1994.
6. Cross, N., Christian, H., and Dorst, K. *Analysing Design Activity*. New York: Wiley, 1996.
7. Dhillon, B. S. *Engineering Design: A Modern Approach*. Toronto: Irwin, 1995.
8. Dieter, G. *Engineering Design*. New York: McGraw-Hill, 1983.
9. Dym, C. L. *Engineering Design*: *A Synthesis of Views*. Cambridge, UK: Cambridge University Press, 1994.
10. Eekels, J., and Rooznburg, N. F. M. "A Methodological Comparison of Structures of Scientific Research and Engineering Design: Their Similarities and Differences," *Design Studies*, Vol. 12, No. 4, pp. 197–203, 1991.
11. Fleddermann, C. B. *Engineering Ethics*. Upper Saddle River, NJ: Prentice Hall, 1999.
12. Hensel, E. "A Multi-Faceted Design Process for Multi-Disciplinary Capstone Design Projects," Proceedings of the 2001 American Society for Engineering Education Annual Conference and Exposition, Albuquerque, NM, 2001.
13. Hill, P. H. *The Science of Engineering Design*. New York: McGraw-Hill, 1983.
14. Horenstein, M. N. *Design Concepts for Engineers*. Upper Saddle River, NJ: Prentice Hall, 1999.
15. Johnson, R. C. *Mechanical Design Synthesis*. Huntington, NY: Krieger, 1978.
16. Watson, S. R. "Civil Engineering History Gives Valuable Lessons," *Civil Engineering*, pp. 48–51, 1975.
17. Jansson, D. G., Condoor, S. S., and Brock, H. R. "Cognition in Design: Viewing the Hidden Side of the Design Process," *Environment and Planning B*, *Planning and Design*, Vol. 19, pp. 257–271, 1993.
18. Karuppoor, S. S., Burger, C. P., and Chona, R. "A Way of Doing Design," Proceedings of the 2001 American Society for Engineering Education Annual Conference and Exposition, Albuquerque, NM, 2001.
19. Kelley D. S., Newcomer, J. L., and McKell, E. K. "The Design Process, Ideation and Computer-Aided Design," Proceedings of the 2001 American Society for Engineering Education Annual Conference and Exposition, Albuquerque, NM, 2001.
20. Pahl, G., and Beitz, W. *Engineering Design: A Systematic Approach*. New York: Springer-Verlag, 1996.
21. Pugh, S. *Total Design*. Reading, MA: Addison-Wesley, 1990.
22. Ray, M. S. *Elements of Engineering Design*. Englewood Cliffs, NJ: Prentice Hall, 1985.
23. Radcliffe, D. F., and Lee, T. Y. "Design Methods used by Undergraduate Students," *Design Studies*, Vol. 10, No. 4, pp. 199–207, 1989.
24. Ross, S. S. *Construction Disasters: Design Failures, Causes and Preventions*. pp. 303–329. New York: McGraw-Hill, 1984.
25. Sickafus, E. N. *Unified Structured Inventive Thinking*. New York: Ntelleck, 1997.
26. Siddall, J. N. "Mechanical Design," *ASME Transactions: Journal of Mechanical Design*, Vol. 101, pp. 674–681, 1979.

27. Suh, N. P. *The Principles of Design.* Oxford, UK: Oxford University Press, 1990.
28. Ulman, D. G. *The Mechanical Design Process.* New York: McGraw-Hill, 1992.
29. Ulrich, K. T., and Eppinger, S. D. *Product Design and Development.* New York: McGraw-Hill, 1995.
30. Vidosic, J. P. *Elements of Engineering Design.* New York: The Ronald Press Co., 1969.
31. Walton, J. *Engineering Design: From Art to Practice.* New York: West Publishing Company, 1991.
32. Whitebeck, C. http://web.mit.edu/ethics/www/essays/probcase.html

2 Design Tools

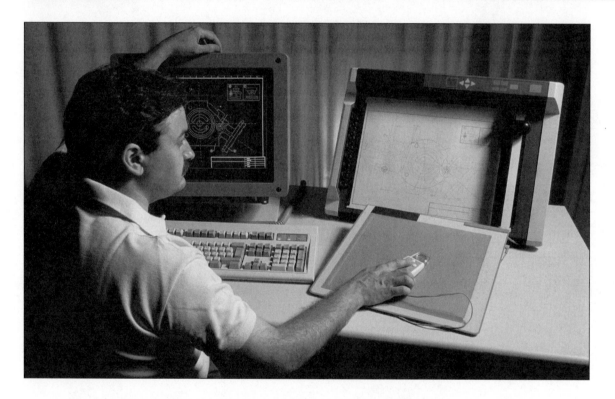

An engineer or practitioner can use many tools to aid in design. Drafting tools as well as computer-assisted design (CAD) programs can help the engineer through the design process steps. (Copyright Kim Steele/Getty Images/PhotoDisc.)

2.1 OBJECTIVES

By the end of this chapter, you should be able to

1. Develop a project schedule utilizing existing tools.
2. Identify and select engineering materials that suit a product.
3. Discuss ergonomic issues as they relate to design.
4. Understand safety considerations.

Before discussing the design process, it is important to state some of the prerequisites of the process, including scheduling, material selection, human factors, and safety issues.

2.2 SCHEDULING

Development of a new product according to the design process is always limited by the time available for the entire process. Over the years, several procedures have been proposed to manage and plan for an artifact. Many projects have failed due to lack of detail. Historically, engineering projects were typically created by one engineer, working alone. The designer, the drafter, and the planner were the same person. In recent years, however, advances in technology require that teams of engineers and technical helpers work beside each other to accomplish a project. In the following subsections, we discuss existing techniques for scheduling.

2.2.1 Gantt Chart

The Gantt chart was introduced by Henry L. Gantt and Frederick Taylor in the early 1900s to facilitate project management. The Gantt chart is in the form of a bar chart and is established as follows:

1. List all events or milestones of the project in an ordered list.
2. Estimate the time required to establish each event.
3. List the starting time and end time for each event.
4. Represent the information in a bar chart.

An example of the Gantt chart is shown in Figure 2.1. Commercial software is available that can help in developing a Gantt chart for the project (e.g., Microsoft Project Manager or Primavera). Many examples of Gantt charts are available on the Web.

2.2.2 CPM/PERT

The critical path method (CPM) and the program evaluation and review technique (PERT), which were developed during the 1950s and 1960s, are the two most widely used approaches. PERT was developed under the guidance of the U.S. Navy Special Projects Office by a team that included members from the Lockheed

Events   Gantt chart

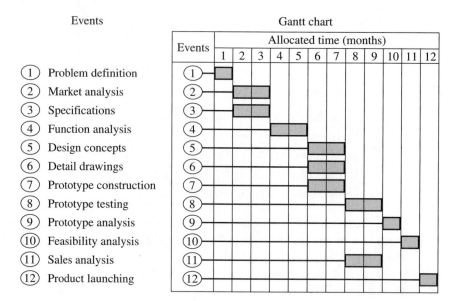

Figure 2.1   Gantt chart.

Missile Systems Division and the consulting firm Booz, Allen and Hamilton. The technique was developed to monitor the efforts of 250 main contractors and 9000 subcontractors who were concerned with the Polaris missile project. CPM was the result of the efforts of the Integrated Engineering Control Group of E.I. duPont de Nemours & Company, and it was developed to monitor activities related to design and construction. A survey of manufacturing companies in the United States revealed that CPM/PERT is used over 65% of the time, among all other methods. A CPM/PERT project generally has the following characteristics:

1. There are clearly defined activities or jobs whose accomplishment results in project completion.
2. Once started, the activity or job continues without interruption.
3. The activities or tasks are independent, which means they may be started, stopped, and performed individually in a prescribed sequence.
4. The activities or jobs are ordered and they follow each other in a specified manner.

Even though the CPM and PERT techniques were developed independently, the basic theory and symbology in both techniques are essentially the same.

2.2.3 Definitions

Several symbols, terms, and definitions are used in developing the CPM/PERT networks:

1. *Event (node):* This represents a point in time in the life of a project. An event can be the beginning or the end of an activity. A circle is used to represent an event. Generally, each network event is identified with a number.

2. *Activity:* This is an effort needed to carry out a certain portion of the project. An activity consumes money, labor, time, and materials and is denoted by a continuous arrow [——→]. The direction of progress on the project is indicated by the arrowhead. Furthermore, an activity both begins and ends at a circle.

3. *Dummy activity:* This is an imaginary activity that takes zero time and no resources to perform. Its purpose is to show appropriate relationships between activities. A dotted arrow is used to represent such an activity.

4. *Network paths:* These are the paths used (or needed) for reaching the project termination point, or the event from the project starting point or event.

5. *Critical path:* This is the longest path with respect to time duration through the PERT/CPM network. In other words, the critical path creates the largest sum of activity duration of all individual network paths.

6. *Earliest event time (EET):* This is the earliest time at which an event occurs, providing all proceeding activities are accomplished within their estimated times.

7. *Latest event time (LET):* This is the latest time at which an event could be reached without delaying the predicted project completion date.

8. *Total float:* This is the latest time of an event, minus the earliest time of the preceding event, and the duration time of the in-between activity.

2.2.4   Network Development

Steps involved in constructing a CPM network are as follows:

1. Break down the design into individual activities and identify each activity.
2. Estimate the time required for each activity.
3. Determine the activity sequence.
4. Construct the CPM network, using the defined symbols.
5. Determine the critical path of the network.

Similarly, the steps involved in developing a PERT network are as follows:

1. Break down the design into individual activities and identify each activity.
2. Determine the activity sequence.
3. Construct the PERT network, using the defined symbols.
4. Obtain the expected time to perform each activity, using the following formula:

$$T_e = \frac{x + 4y + z}{6}$$

where

T_e is the expected time for the activity.

x is the optimistic time estimate for the activity.

y is the most likely time estimate for the activity.

z is the pessimistic time estimate for the activity.

TABLE 2.1   Probability Table

w	Probability
−3	0.0013
−2.5	0.006
−2	0.023
−1.5	0.067
−1	.159
−0.5	0.309
0	.5
0.5	0.69
1	0.84
1.5	0.933
2	0.977
2.5	0.994
3	0.999

5. Determine the network critical path.
6. Compute the variance associated with the estimated expected time of T_e of each activity using the following formula:

$$S^2 = \left(\frac{z-x}{6}\right)^2$$

7. Obtain the probability of accomplishing the design project on the stated date using the following formula:

$$w = \frac{T-T_L}{[\Sigma S_{cr}^2]^{\frac{1}{2}}}$$

where

S_{cr}^2 is the variance of activities on the critical path.

T_L is the last activity's earliest expected completion time, as calculated through the network.

T is the design project due date, expressed in time units. Table 2.1 presents the probabilities for selective values of w.

Example

A mechanical design project was broken down into a number of major jobs or activities, as shown in Table 2.2. A CPM network was developed using the data given in the table, and the critical path of the network was determined, along with the expected project duration time period.

Using the defined symbols and the data given in Table 2.2, the CPM network is shown in Figure 2.2.

TABLE 2.2 Design Project

Activity Description	Activity Identification	Immediate Predecessor Activity or Activities	Activity Duration (Week or Weeks)
Needs and goals Market analysis	A	—	1
Function analysis	B	A	2
Specifications	C	B	1
Alternatives	D	C	4
Evaluations	E	D	3
Prototyping	F	D	4
Analysis	G	E,F	2
	H	G	2
Manufacturing	I	G	4
Marketing	J	H,I,3	

Figure 2.2 CPM chart.

The paths originated at event 1 and terminated at 11 are as follows:

1. A-B-C-D-E-G-I-J time = 20
2. A-B-C-D-F-G-I-J time = 21
3. A-B-C-D-E-G-H-J time = 18
4. A-B-C-D-F-G-H-J time = 19

These results indicate that the longest path is (A-B-C-D-F-G-I-J), and the predicted time for the design project is 21 weeks.

2.3 MATERIAL SELECTION

The design, even of the simplest element, requires the selection of a suitable material and a decision regarding the methods of manufacturing to be used in producing the element. These two factors are closely related, and the choice will affect the shape, appearance, cost, and so on. It may also determine the difference between a commercial success and a commercial failure. As the design becomes more complex and involves more elements, the selection of suitable materials and methods of production becomes more difficult. The design engineer must be sufficiently familiar with the characteristics and properties of the materials and the way in which they can be shaped to ensure that his or her decisions are well made.

2.3.1 Material Classifications and Properties

The various types of materials that may be used in a product design are classified as follows:

1. *Metals:* This important materials classification can be further divided into ferrous and nonferrous alloys. Ferrous alloys are based on iron, which has mechanical properties. Nonferrous alloys are based on materials other than iron, such as copper, tin, aluminum, and lead.

2. *Ceramics and glass:* These are the result of the combination of metallic and nonmetallic elements. They are good insulators, brittle, thermally stable, and more wear resistant compared with metals. They are also harder and lower in thermal expansion than most metals.

3. *Woods and organics:* These are obtained from trees and plants. Their major advantage is that they are a renewable resource. However, some of their drawbacks include that they absorb water, require special treatment to prevent rotting, and are more flammable.

4. *Polymers or plastics:* These materials change viscosity with variation in temperature; therefore, they are easy to mold into a given shape. There are several benefits of polymers: They are good insulators; they are resistant to chemicals and water; they have a smooth surface finish; and they are available in many colors, which eliminates the need for painting. There are several drawbacks to polymers, however: They are low in strength; they deteriorate in ultraviolet light; and they have excessive creep at all temperatures.

The properties of materials can be divided into four categories: mechanical, thermal, chemical, and fabrication. *Mechanical properties* include fatigue, strength, wear, hardness, and plasticity. Thermal properties include absorptivity, fire resistance, and expansion coefficient. *Physical properties* include permeability, viscosity, crystal structure, and porosity. *Chemical properties* include corrosion, oxidation, and hydraulic permeability. Four important electrical properties are hysteresis, conductivity, coercive force, and dielectric constant. There are several fabrication properties, including weldability, castability, and heat treatability. A partial list of selected material properties is provided in Figure 2.3.

2.3.2 Material Selection Process

Several systematic approaches have been developed for selecting materials; one of these is listed here:

1. *Perform a material requirements analysis:* This step is concerned with determining environmental and service conditions under which the product will have to operate.

2. *List the suitable materials:* This step calls for filtering through the available materials and obtaining several suitable candidates.

3. *Choose the most suitable material:* This step involves analysis of the material listed, with respect to such factors as cost, performance, availability, and fabrication ability, and then selecting the most suitable material.

Material	Density lb/in^3	ν Poisson's Ratio	E 10^4 psi	σ_y 10^3 psi	σ_{ult} 10^3 psi	Melting Temperature °F	Thermal Conductivity lb/sec °F
Pure Metals							
Berylium	0.066		44	55	90	2340	19.9
Copper	0.32	0.33	17	10	32	1980	51
Lead	0.41	0.43	2	2	2.5	621	4.5
Nickel	0.32		30	20	70	2625	7.9
Tungsten	0.70		50		300	6092	26.2
Aluminum	0.1	0.33	10	3.5	11	1200	29.0
Alloys							
Aluminum 2024-T4	0.1	0.33	10.6	44	60	1075	15.8
Brass	0.31	0.35	15	60	74	1710	14.9
Cast iron (25T)	0.26	0.2	13	24	120	2150	5.8
Steel: 0.2% C							
Hot Rolled	0.283	0.27	30	40	70	2760	6.5
Cold Rolled	0.283	0.27	30	65	80	2760	6.5
Stainless Steel							
Type 302 C.R.	0.286	0.3	29	100	140	2575	1.9
Ceramics							
Crystalline Glass	0.09	0.25	12.5	20		2280	0.24
Fused Silica Glass	0.08	0.17	10.5			2880	0.17
Plastics							
Cellulose Acetate	0.047	0.4	0.25	5	20		0.032
Nylon	0.041	0.4	0.41	8	13		0.03
Epoxy	0.04		0.65	7	30		0.10

Figure 2.3   Material properties.

4. *Obtain the necessary test data:* This step calls for determining the important properties of the selected material, experimentally under real-life operational conditions.

5. *Product specification fulfillment:* The selected material must satisfy the stated specifications.

6. *Cost:* This important factor plays a dominant role in the marketing of the end product.

7. *Material availability:* The selected material must be available at a reasonable cost.

8. *Material joining approach:* Under real-life conditions, it may be impractical to produce an element using a single piece of material. This may call for manufacturing the components with different pieces of material joined together to form a single unit.

9. *Fabrication:* Engineering products usually require some level of fabrication, and different fabrication techniques are available. Factors affecting

the fabrication method selected include time constraints, material type, product application, cost, and quantity to be manufactured.

10. *Technical issues:* Technical factors are mainly concerned with the material mechanical properties. Examples include strength compared with anticipated load, safety factors, temperature variation, and potential loading changes.

2.3.3 Primary Manufacturing Methods

The primary manufacturing methods, which are used to convert a material into the basic shape required, are as follows:

1. *Casting:* Casting is a widely used first step in the manufacturing process. During casting, an item takes on its initial usable shape. In the casting process, a solid is melted down and heated to a desirable temperature level. The molten material is then poured into a mold made in the required shape. The cast items may range in size and weight from a fraction of an inch to several yards. Typical examples include a zipper's individual teeth, and the stern frames of ships.

2. *Forging:* Forging, which is among the most important methods of manufacturing items of high-performance uses, involves changing the shape of the piece of material by exerting force on that piece. The methods of applying pressure include the mechanical press, hydraulic press, and drop hammer. Products such as crankshafts, wrenches, and connecting rods are the result of forging. The material to be forged may be hot or cold.

3. *Machining:* Machining involves removing unwanted material from a block of material according to given specifications, such as size, shape, and finish, to produce a final product. There are many machining processes, such as milling, boring, grinding, and drilling.

4. *Welding:* Welding is a versatile production process that is used for combining items produced by some other means of manufacturing. Welding is the process of permanently joining two materials through coalescence, which involves a combination of pressure and surface conditions.

2.4 HUMAN FACTORS

The science that considers humans and their reactions to familiar, as well as strange, environments is known by several terms, including *human engineering*, human performance, and *ergonomics*. Adapting a human to fit a given environment is almost impossible, so methods for adapting the environment to suit the human must be devised. Human behavior is critical to the success of an engineering system. Therefore, it is important to consider typical human behaviors during the design phase. Examples of typical human behaviors include the following:

1. People are often reluctant to admit errors.
2. People usually perform tasks while thinking about other things.
3. People frequently misread or overlook instructions and labels.
4. People often respond irrationally in emergency situations.

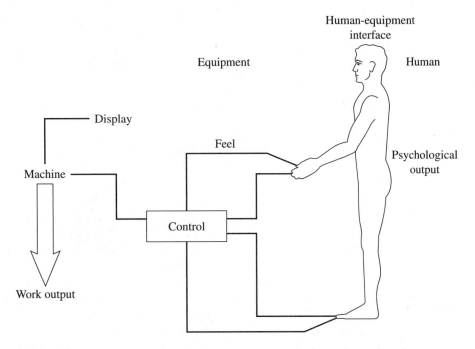

Figure 2.4   Human–machine system.

5. A significant percentage of people become complacent after successfully handling dangerous items over a long period of time.

6. Most people fail to recheck outlined procedures for errors.

7. People are generally poor estimators of speed, clearance, or distance. They frequently overestimate short distance and underestimate large distance.

8. People are generally too impatient to take the time needed to observe.

9. People are often reluctant to admit that they cannot see objects well, due either to poor eyesight or to inadequate illumination. They generally use their hands for examining or testing.

Every time people dial telephones, drive cars, study oscilloscopes, use computers, or form parts on a lathe, they have joined their sensing, decision-making, and muscular powers into an engineering system. Figure 2.4 shows the coupling of components that are necessary to consider a human as an integral part of an efficient human–machine system.

2.4.1   Human Sensory Capabilities

Human sensory capabilities include

1. *Sight:* Human eyes see differently from different angles or positions. For example, looking straight ahead, the eyes can perceive all colors. However, with an increase in the viewing angle, human perception begins to decrease. The limits of color vision are given in Table 2.3. Also, at night or in the dark, small-sized orange, blue, green, and yellow lights are impossible to distinguish from a distance.

TABLE 2.3	Human Color Vision					
Situation	**Green**	**Blue**	**Yellow**	**White**	**Green-Red**	**Red**
Vertical	40°	80°	95°	130°	—	45°
Horizontal	—	100°	120°	180°	60°	—

2. *Noise:* The performance quality of a task requiring intense concentration can be affected by noise. It is an established fact that noise contributes to people's feelings, such as irritability, boredom, and well-being. A noise level of 90 dB is considered harmless, while above 100 dB is unsafe. Levels in excess of 130 dB are considered unpleasant and may actually become harmful.

3. *Touch:* Touch adds to, or may even replace, the information transmitted to the brain by eyes and ears. For example, it is possible to distinguish knob shapes with touch alone. This ability could be valuable if, for instance, the power goes out and there is no light.

4. *Vibration and motion:* It is an accepted fact that poor performance of physical and mental tasks by equipment operators could be partially or fully attributed to vibrations. For example, eye strain, headaches, and motion sickness could result from low-frequency, large-amplitude vibrations. Useful guidelines for the effects of vibration and motion are as follows:
   a. Eliminate vibrations with an amplitude greater than 0.08 mm.
   b. Use such devices as shock absorbers and springs wherever possible.
   c. Use damping materials or cushioned seats to reduce vibration wherever possible.
   d. Seated humans are most affected by vertical vibration, so use this information to reduce vertical vibrations.
   e. The resonant frequency of the human vertical trunk, in the seated position, is between 3 and 4 cycles per second. Therefore, avoid any seating that would result in or would transmit vibrations of 3 to 4 cycles per second.

2.4.2 Anthropometric Data

Anthropometry is the science that deals with the dimensions of the human body. Anthropometric data are divided into statistical groups known as percentiles. If 100 people are lined up from smallest to largest in any given respect, they would be classified from the 1 percentile to the 100 percentile. The 2.5 percentile means that designs based on this series of dimensions would allow up to 2.5% of the population to use the system and the remaining 97.5% would be excluded. Figure 2.5 shows the dimensions of a standing adult male.

2.5 SAFETY CONSIDERATIONS

One very important consideration in engineering design maintains that the resulting product is safe for humans. In recent years, the liability decisions made by the courts further increased the importance of safety. According to the National Safety

Figure 2.5   Standing adult male.

Council, over 105,000 Americans were accidentally killed and over 10 million suffered a disabling injury in 1980, which translated into $83.2 billion in disability payments. Safety is hardly new; it has been important for thousands of years. There are many safety functions used during the design process, including

1. Developing accident prevention requirements for the basic design
2. Participating in the design reviews
3. Performing hazard analyses during the product design cycle

2.5.1   Safety Analysis Techniques

The two techniques used for design stage and reliability analysis—failure modes and effect analysis (FMEA) and fault trees—can also be applied for safety analyses.

Failure modes and effect analysis. This method was originally developed for use in the design and development of flight control systems. The method can also be used to evaluate design at the initial stage from the point of view of safety. Basically, the technique calls for listing the potential failure modes of each part, as well as the effects on the parts and on humans. The technique may be broken down into seven steps:

1. Defining system boundaries and requirements
2. Listing all items
3. Identifying each component and its associated failure modes
4. Assigning an occurrence probability or failure rate to each failure mode
5. Listing the effects of each failure mode on concerned items and people
6. Entering remarks for each possible failure mode
7. Reviewing and initiating appropriate corrective measures

Fault trees. This technique uses various symbols. It starts by identifying an undesirable event, called the top event, and then successively asking, "How could this event occur?" This process continues until the fault events do not require further development. If occurrence data are known for the basic or primary fault event, the occurrence measure for the top event can be calculated.

2.6   TECHNICAL WRITING AND PRESENTATION

In the engineering and scientific professions, communication skills are as important as in other fields. Following the development of an innovative design and a cost estimate that predicts large profit, the designer must be able to communicate the findings to other people. An old adage reminds us that a tree falling in a forest doesn't make a sound unless there is someone to listen. Similarly, the best technical design in the world might never be implemented unless the designer can communicate the design to the proper people, in the right way.

The quality of a report generally provides an image in the reader's mind that, in large measure, determines the reader's impression of the quality of work. Of course, an excellent job of report writing cannot disguise a sloppy investigation, but many excellent design studies have not received proper attention and credit because the work was reported in a careless manner. A formal technical report is usually written at the end of a project. Generally, it is a complete, stand-alone document written for people who have widely diverse backgrounds. Therefore, much more detail is required. The outline of a typical formal report is as follows:

1. Cover page:
   a. Title of the report
   b. Name of author(s)
   c. Address
2. Summary: Summarizes the work, and should also include a short conclusion.
3. Table of contents.
4. List of figures and list of tables, including the corresponding page number.
5. Introduction: Background of the work (market analysis) to acquaint the reader with the problems with and purposes of carrying on the work.
6. Design process: Details of the procedure followed in the design process.
7. Discussion: This section should contain the comprehensive explanation of the results. The discussion may be divided into several subsections, such as
   a. Technical analysis (force requirement, speed, etc.)
   b. Details of the build up of the artifact
   c. Equipment used
   d. Assembly and setup procedures
   e. Experimental setups and results
   f. Details of the final results
8. Conclusions: This section states in as concise a form as possible the conclusions that can be drawn from the study.
9. References: This section lists all the documents to which the writers referred. The information on each document must be complete and must follow the same format throughout.
10. Appendices: The appendices include material deemed beyond the scope of the main body of the report. The appendix section may be divided into as many subsections as necessary.

2.6.1   Steps in Writing a Report

The five operations involved in the writing of a high-quality report are best remembered with the acronym POWER:

1. **P**: Plan the writing
2. **O**: Outline the report
3. **W**: Write
4. **E**: Edit
5. **R**: Rewrite

Good writers usually produce good reports. Some of the attributes of good writing are as follows:

1. Write as objectively as possible. Do not become emotionally involved or attached to a problem or a solution.

2. Be reasonably methodical.
3. Record whatever is learned, and keep in mind that whatever work is performed must eventually be documented.
4. Always strive for clarity in writing, and keep in mind that the written material should be simple and straightforward.
5. Deliver the written material on time.

There are also certain qualities associated with good reports, such as the following:

1. Delivered on the due date
2. Effectively answer readers' questions as they arise
3. Give a good first impression
4. Read coherently
5. Contain an effective summary and conclusion
6. Are written clearly and concisely, and avoid vague or superfluous phrases
7. Provide pertinent information

2.6.2   Illustration Guidelines

Visual elements, such as figures, charts, and graphs, in technical reports have specific purposes and convey specific information. Visuals are used to explain, illustrate, demonstrate, verify, or support written material. Visuals are only valuable if their presentation is effective. The general guidelines for preparing effective visuals (illustrations) are as follows:

1. Reference all illustrations in the text.
2. Reference the data source.
3. Carefully plan the placement of illustrations.
4. Specify all units of measure and the scale used in the drawing.
5. Label each illustration with an identifying caption and title. Include figure source, if the figure was obtained from another document.
6. Spell words out rather than using abbreviations.
7. When a document has five or more figures, include a list of figures at the beginning of the report.
8. Avoid putting too much data in an illustration.

2.6.3   Mechanics of Writing

The following suggestions will help you avoid some of the most common mistakes in your writing:

- *Paragraph structure:* Each paragraph should begin with a topic sentence that provides an overall understanding of the paragraph. Each paragraph should have a single theme or conclusion, and the topic sentence states that theme or conclusion.
- *Sentence length:* Sentences should be kept as short as possible so that their structure is simple and readable. Long sentences require complex

construction, provide an abundance of opportunity for grammatical errors, take considerable writing time, and slow the reader. Long sentences are often the result of putting together two independent thoughts that could be stated better in separate sentences.

- *Pronouns:* There is no room for ambiguity between a pronoun and the noun for which it is used. Novices commonly use *it*, *this*, *that*, and so on where it would be better to use one of several nouns. It may be clear to the writer but is often ambiguous to the reader. In general, personal pronouns (*I, you, he, she, my, mine, our, us*) are not used in reports.
- *Spelling and punctuation:* Errors in these basic elements of writing are inexcusable in the final draft of the report.
- *Tense:* Use the following rules when choosing the verb tense:
  a. *Past tense:* Use to describe work done when you were building or designing, or in general when referring to past events.
  b. *Present tense:* Use in reference to items and ideas in the report itself.
  c. *Future tense:* Use in making predictions from the data or results that will be applicable in the future.

2.7 PRESENTATION STYLE

Unlike advertising executives, engineers are ill equipped to sell their ideas. Second-hand information represented by company officials may not answer all of the client's questions. This section discusses the specifics and techniques of oral presentations.

2.7.1 Objective

Every presentation should have an objective. The speaker's main objective is to deliver the message (objective) to the audience. The objectives may vary from one presentation to another. To identify the real objective, ask the following question: "If everything goes perfectly, what do I intend to achieve?" Realize who your audience is and what their educational level may be.

In most cases, presentation time is limited. It is of utmost importance to keep within the scheduled time during question/answer sessions. To stay within time restraints, detailed planning is required. Different tools can be used in the presentation, such as slides, models, transparencies, audiovisual, and the Web. Make sure to account for the time needed to shift from one medium to another.

An easy way to evaluate the effectiveness of a presentation while on a team is to practice through role-playing. One person can play the role of the speaker, and the rest of the team can act as the audience, and possibly even play the devil's advocate. In this way, the team gains valuable experience before attempting the actual presentation.

2.7.2 Oral Presentation Obstacles

To sell your ideas to others, you should first be convinced that your ideas will accomplish the task at hand. Oral presentation requires a high degree of creativity.

People resist change, although they may announce that they embrace it. Humans like familiar methods. Change requires additional efforts, which humans, in general, resist. Typical reactions to change include the following:

- We tried before.
- It is a too radical change.
- We have never done that before.
- Get back to reality.
- We have always done it this way.
- I don't like the idea.

Keep in mind the following:

- Only 70% of the spoken word is actually received and understood. Complete understanding can come through repetition and redundancy in speech.
- People mostly understand three-dimensional objects. Two-dimensional projections need to be transmitted with added details.
- People usually perceive problems from their own perspectives.
- Convey ideas so that they may be interpreted with the least expenditure of energy.

2.7.3   Oral Presentation Dos and Don'ts

Remember the following in making oral presentations:

1. Know your audience thoroughly.
2. Never read solely from notes, a sheet, or directly from an overhead projector. You can use your notes for reference, but remember to make eye contact with your audience from time to time.
3. Bring the audience up to the speed in the first few moments.
4. Stay within the time allotted.
5. Include relevant humorous stories, anecdotes, or jokes (only if you are good at it).
6. Avoid using specialized technical jargon. Explain the terms you feel the audience may not know.
7. Understand your message clearly. The entire goal is to communicate the message clearly.
8. Practice, practice, practice! You may like to memorize the introduction and concluding remarks.
9. The dry run is a dress rehearsal. Use it to iron out problems in delivery, organization, and timing.
10. Avoid mannerisms: Speak confidently but not aggressively.
11. Maintain eye contact with audience members, and keep shifting that contact through the talk.
12. Never talk to the board or to empty space.

13. Present the material in a clever fashion, but not in a cheap and sensational fashion. Be genuinely sincere and professional.

14. A logical presentation is much more critical in an oral one than it is in a written one.

2.7.4 Oral Presentation Techniques

The following will help you make your oral presentation as effective as possible:

1. Visual aids (sketches, graphs, drawings, photos, models, slides, transparencies, and the Web) often convey information more efficiently and effectively. Visual aids permit the use of both the hearing and seeing senses, and they help the speaker.

2. Limit slides to not more than one per minute.

3. Each slide should contain one idea.

4. The first slide should show the title of your presentation and the names of the collaborators.

5. The second slide should give a brief outline of the presentation.

6. The last slide should summarize the message you just delivered.

7. If you need to show a slide more than once, use a second copy.

8. Avoid leaving a slide on the screen if you have finished discussion on the topic.

9. Never read directly from the slide. Spoken words should complement the slides. Prepare notes for each slide and use them during practice.

10. Use graphs to explain variations. Clearly label the axis, data, and title. Acknowledge the source.

11. Every graph should have a message (idea). Color should enhance the communication, not distract from it.

12. Audiences respond to well-organized information. That means
    a. Efficient presentation
    b. All assumptions clearly stated and justified
    c. Sources of information and facts clearly outlined

13. Begin with the presentation of the problem and conclusion/recommendation (primary goal).

14. Finish ahead of schedule and be prepared for the question/answer session.

2.7.5 Question/Answer Session

The question/answer session is very important. It shows the enthusiasm of the audience and usually reveals interest and attention. In the Q/A session you should

1. Allow the questioner to complete the question before answering.

2. Avoid being argumentative.

3. Do not let the questioner feel that the question is stupid.

4. Adjourn the meeting if the questions slack off.

5. Thank the audience one final time after the Q/A session.

2.8   PROBLEMS

2.8.1   Team Activities

Scheduling

1. Painting a two-story house was broken down into a number of major jobs or activities, as shown in Table 2.4.
   (a) Develop a CPM network.
   (b) Determine the critical path of the network.
   (c) Determine the expected project duration time period.

TABLE 2.4   Team Problem 1

Activity	Identification	Predecessor	Duration
Contract signed	A	—	2
Purchase of material	B	A	2
Ladder & staging in site	C	A	2
Preparation of surface	D	C,B	5
Base coat complete	E	D	6
Base coat inspected	F	E	2
Trim coat complete	G	E	5
Trim coat inspected	H	G,F	2
Final inspection	I	H	2
Removal of staging	J	H	2
Final cleanup	K	I,J	2

2. Assume that the optimistic, most likely, and pessimistic activity times are as given in Table 2.5.
   (a) Complete the table.
   (b) Find the probability of finishing the job in 32 weeks.

TABLE 2.5   Team Problem 2

Pessimistic	Most Likely	Optimistic	Expected Time	Variance
3	2	1		
3	2	1		
3	2	1		
7	5	3		
9	6	4		
3	2	1		
7	5	4		
3	2	1		
3	2	1		
3	2	1		
3	2	1		

3. You should work on this activity during lab hours.
   (a) Develop a CPM network and determine the critical path for the events defined in the Gantt chart in Figure 2.1.
   (b) Develop a Gantt chart for the events defined in the Table 2.2.
   (c) For the events defined in Table 2.6,
       i. Complete the table.
       ii. Draw the PERT network.
       iii. Find the probability of finishing the task on time if the design due date is after 95 days.

Material Selection

4. In two opposite columns, list five uses of metals that are not appropriate for plastics and five uses of plastics not appropriate for metals.
5. From the table of mechanical properties (see Figure 2.3) of selected materials, list the differences between pure metals and metal alloys.

TABLE 2.6	Team Problem 3				
Predecessor	**Optimistic**	**Most Likely**	**Pessimistic**	**Expected**	**Variance**
1	3	5	8		
1	4	6	9		
3	3	4	5		
2	2	3	4		
5	3	4	5		
1	8	12	14		
4	14	18	21		
7	5	10	14		
7	5	10	14		
7	5	10	14		
7	5	10	14		
7	5	10	14		
12	4	6	10		
11	4	6	10		
10	4	6	10		
9	4	6	10		
8	4	6	10		
13	10	12	18		
6	16	18	24		
14	7	10	15		
14	10	15	22		
17	5	9	9		
18	4	6	8		
16	6	8	12		
15	3	8	12		
19	3	4	5		

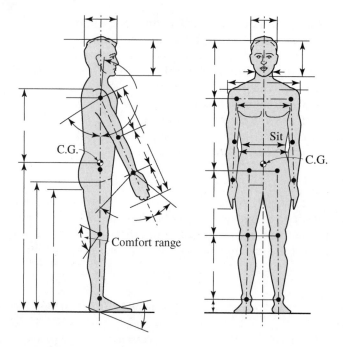

Figure 2.6   Anthropometric chart.

6. Based on the table, compare between metals-alloys/ceramics/plastics in terms of weight, strength, stiffness, and thermal expansion.
7. What is the difference between forging, casting, and welding?
8. Choose the captain of your team to measure and record his or her anthropomorphic dimensions on the chart shown in Figure 2.6. In what percentile would you classify your captain?

2.8.2   Individual Activities

1. Define the following terms:
   (a) CPM
   (b) Gantt chart
   (c) PERT
2. Discuss four major classifications of materials.
3. List the classifications of manufacturing processes.
4. Describe the following:
   (a) Casting
   (b) Forging
   (c) Welding
5. Assume that an engineering course term project is broken down into a number of major jobs or activities, as shown in Table 2.7.
   (a) Draw a Gantt chart.
   (b) Develop a CPM network.
   (c) Determine the critical path of the network.
   (d) Determine the project duration time period.

TABLE 2.7   Term Project

Activity Description	Activity Identification	Immediate Predecessor Activity	Activity Duration (Days)
Literature collection	A	—	7
Literature review	B	A	4
Outline preparation	C	B	1
Analysis	D	B	10
Report writing	E	C	5
Typing	F	E	3
Revision	G	E	4
Final draft	H	G	2

6. Assume that, for example, the optimistic, most likely, and pessimistic activity times are as in Table 2.8. Calculate each activity's expected time and variance, and the probability of accomplishing the design project in 32.5 weeks. In addition, calculate each event's earliest and latest event times.

7. Describe the following terms:
   (a) Ergonomics
   (b) Human error
   (c) Human reliability
   (d) Human engineering
       Hint: Search in the library for more information.

8. Consider that you have been assigned the task of designing the environment for an industrial worker who must be seated at a control console for four continuous hours each day. The worker monitors and controls plant and laboratory temperature and humidity. Describe (list) the types of environment that would be considered ideal for this operator.

9. Design an optimum display of controls, indicators, and utility items listed in the following proposed dash layout (see also Figure 2.7), paying

TABLE 2.8   Activity Time Estimates (Weeks)

Pessimistic	Most Likely	Optimistic	Expected Time	Variance
4	2	1	2.16	0.25
3	2	1	2	0.11
4	2	1	2.16	0.25
8	5	3	5.16	0.7
5	4	2	3.83	0.25
7	4	3	4.33	0.44
5	3	2	3.17	0.25
5	3	2	3.17	0.25
9	6	3	6	1
7	4	2	4.17	0.69

Figure 2.7 Proposed dash design layout.

particular attention to human engineering principles. Where present dials and indicators are ineffective, redesign the display of information for efficient use by the human operator. Ask yourself the following questions about the information to be conveyed: "Which is the most important?" "Least important?" "Which is the most often used?" "Least often used?" "What information must the indicator convey to the operator?" "Does it provide this information?" "Does it overinform?" "Underinform?" "What medium does it use?" Most indicators are used visually, and eyes are often overburdened in their functional usage. This would be critical in a high-speed car. Could the same information be obtained by other means, such as light, sound, or touch?

	Indicators	**Utility Items**
	Speedometer	Glove compartment
	Odometer	Radio
	Fuel	Cigarette lighter
	Engine temperature	Ventilation
Signals	Battery	Rear view mirror
	Oil pressure	
(Starter)	High-beam lights	

10. What is meant by each of the following terms?
    (a) Elastic limit
    (b) Yield point
    (c) Ultimate strength
    (d) Stiffness
11. What is the major requirement of a material if minimum deflection is required? Give an example.
12. What is meant by ductility? Suggest a method to measure ductility.

2.9 SELECTED BIBLIOGRAPHY

1. Beard, P. D., and Talbot, T. F. "What Determines If a Design Is Safe?," Proceedings of ASME Winter Annual Meeting, pp. 90–WA/DE-20, New York, 1990.
2. Blake, A. *Practical Stress Analysis in Engineering Design*. New York: Marcel Dekker, 1982.
3. Burgess, J. H. *Designing for Humans: The Human Factor in Engineering*. Princeton, NJ: Petrocelli Books, 1986.
4. Burr, A. C. *Mechanical Analysis and Design*. New York: Elsevier Science, 1962.
5. Collins, J. A. *Failure of Materials in Mechanical Design*. New York: Wiley, 1981.
6. Cook, N. H. *Mechanics and Materials for Design*. New York: McGraw-Hill, 1984.
7. Cullum, R. D. *Handbook of Engineering Design*. London: Butterworth, 1988.
8. Deutschman, A. D. *Machine Design: Theory and Practice*. New York: Macmillan, 1975.
9. Dhillon, B. S. *Engineering Design: A Modern Approach*. Toronto: Irwin, 1996.
10. Diesch, K. H. *Analytical Methods in Project Management*. Ames, Iowa: Iowa State University, 1987.
11. Dieter, G. E. *Engineering Design: A Material and Processing Approach*. New York: McGraw-Hill, 1983.
12. Doyle, L. E. *Manufacturing Processes and Material for Engineers*. Englewood Cliffs, NJ: Prentice Hall, 1985.
13. Dreyfuss, H. *The Measure of Man: Human Factors in Design*. New York: Whitney Library of Design, 1967.
14. Frederick, S. W. "Human Energy in Manual Lifting," *Modern Materials Handling*, Vol. 14, pp. 74–76, 1959.
15. Furman, T. T. *Approximate Methods in Engineering Design*. New York: Academic Press, 1980.
16. Glegg, G. L. *The Development of Design*. Cambridge, UK: Cambridge University Press, 1981.
17. Grandjean, E. *Fitting the Task of the Man: An Ergonomic Approach*. London: Taylor and Francis, 1980.
18. Greenwood, D. C. *Engineering Data for Product Design*. New York: McGraw-Hill, 1961.
19. Hajek, V. *Management of Engineering Projects*. New York: McGraw-Hill, 1984.
20. Hilton, J. R. *Design Engineering Project Management: A Reference*. Lancaster, PA: Technomic Publishing Co., 1985.
21. Hindhede, A. *Machine Design Fundamentals*. New York: Wiley, 1983.
22. Hubka, V., and Eder, W. E. *Principles of Engineering Design*. London: Butterworth Scientific, 1982.
23. Lupton, T. *Human Factors: Man, Machine and New Technology*. New York: Springer-Verlag, 1986.
24. McCormick, E. J., and Sanders, M. S. *Human Factors in Engineering and Design*. New York: McGraw-Hill, 1982.

25. Meredith, D. D. *Design and Planning of Engineering Systems.* Englewood Cliffs, NJ: Prentice Hall, 1985.
26. Oborne, D. J. *Ergonomics at Work.* New York: Wiley, 1982.
27. Papanek, V. *Design for Human Scale.* New York: Van Nostrand Reinhold, 1983.
28. Pheasant, S. T. *Bodyspace: Anthropometry, Ergonomics and Design.* London: Taylor and Francis, 1986.
29. Schmidtke, H. *Ergonomic Data for Equipment Design.* New York: Plenum Press, 1985.
30. Shigley, J. E., and Mischke, C. R. *Mechanical Engineering Design.* New York: McGraw-Hill, 1983.
31. Woodson, W. E. *Human Factors Design Handbook.* New York: McGraw-Hill, 1981.

3 Needs and Goals

Before a detailed product, such as this dashboard, can be designed, a list of needs and goals for such a product should be formulated by the design team. (Courtesy Daimler-Chrysler.)

3.1 OBJECTIVES

By the end of this chapter, you should be able to

1. Identify and abstract the statement of need.
2. Develop an objective tree for the need statement.

In general, need statements are not well defined. Even if they are, you will always need to make sure that what you are developing is what the customer needs. The mechanism of clarifying the need statement is discussed in this chapter and the following three chapters. This chapter discusses organizing customer needs to match your understanding. Chapter 4 discusses clarifying the customer need by searching what is already on the market, and who you are competing with. Chapter 5 provides the tools to help you shape your ideas to solve the problem. Chapter 6 discusses revising objectives from qualitative definitions into quantitative definitions. In Chapter 7, through this systematic process, you begin to develop definitive solutions.

3.2 PROBLEM DEFINITION: NEED STATEMENT

During the course of human development, different kinds of needs existed. For instance, there has always been and always will be a need for improving and making new designs. Lincoln Steffenes wrote, "The world is yours, nothing is done, nothing is known. The greatest poem isn't written, the best railroad isn't build yet, the perfect state hasn't been thought of. Everything remains to be done right, everything." The engineer is a person who applies scientific knowledge to satisfy humankind's needs. It should be emphasized that the ability to design is a characteristic of an engineer.

One serious difficulty that engineers must overcome deals with the form in which problems are often presented to them. Even if some goals are given to the engineer, they are often not specifically stated. Problems may be presented as "The shaft is breaking." "The controls aren't producing the desired effect", and "It costs too much to operate this engine." Thus, the first task of the engineer involves determining the real problems. Then, the engineer must determine the extent and confines of the goals.

It is necessary to formulate a clear, exact statement of the problem in engineering words and symbols. It is also necessary to isolate the problem from the general situation and to delineate its form. This definition should clearly identify every aspect of the problem on which attention should be concentrated. The nonessential should be stripped away, and the individual characteristics of the problem should be differentiated. It should be determined whether or not the immediate problem is part of the larger problem. If it is, its relationship to the total part should be determined.

A designer is presented with a situation involving waste of irrigation water in public parks. The park keepers forget to turn off the water. A general formulation of the problem would be, "What can we do to minimize the possibility of workers forgetting to turn off the water before the end of their shift?" An engineer would ask questions such as, "Why do workers continue to forget to turn off the water?", "What is the sequence of events that workers use during their daily activities?", "What will happen if a keeper does not show up for his/her shift?", "Do we need to manually turn on and off the water?" A more precise form of the problem statement would be, "How do we prevent irrigation water waste in public parks?"

A company has proposed to use the density gradient to isolate red blood cells from whole blood and thus to treat white blood cells with a light-activated drug. The designer should ask questions such as, "Is it necessary to use the density gradient if other methods of separation would be capable of isolating the red blood cells from the whole blood?", "If the white cells are being treated, why don't we isolate the white cells from the whole blood rather than isolate red blood cells?", and "Why don't we impede the light into the blood and reduce the need for separation?" Designers need to abstract the need statement from its current state to a statement that they can base their design on. Vague statements from the customer usually result in bad design.

Consider the following example. An engineer is presented with a problem caused by formation of ice on roofs. The ice forms during certain types of weather, falls away from the roofs, and causes damage to vehicles and people below. A general formulation of this problem might be, "How do we prevent ice from forming on roofs?" By asking questions as, "What would happen if ice did form?", "What will cause the ice to fall?", and "What harm would such formation do?" it was determined that the first definition was much too narrow. A much broader definition was, "How do we prevent ice that forms on roofs from doing harm or damage to people and equipment below?"

Before an engineer can define the problem properly, he or she must recognize all the problems that exist. Most of the failures in machines do not occur because we make mistakes in analyzing the problem, but because we fail to recognize that there is a problem. For example, the first-year maintenance cost for a paint drying oven was $2000 (an oven whose original cost is $1700). The high cost was due to the frequent dismantling of oil in bronze sleeve bearings mounted within the baking chamber. Once the problem was recognized, the solution was relatively simple (mount the bearings outside the oven).

A motor is used to drive a magnetically driven pump. Powerful magnets are attached at the shaft of the motor. The motor continues to fail because the magnets continue to interact with the stator magnets in the motor. Once the problem was identified, the solution was easy (put the magnets at a further location from the motor; you can use a belt system to drive the magnets).

3.3   OBJECTIVE TREE

An important step in design is clarifying the design objectives. This step is helpful in all design stages. The objective statement

1. Should be easy to understand by the client and the designer.
2. Should be agreed on by the client and the designer.
3. Controls and manages the design.

The objective tree method offers a clear and useful format for the objective statement. It shows in a diagrammatic form the ways in which different objectives are related to each other. The objective tree method procedure was summarized by Cross as follows:

1. Prepare a list of design objectives. These take the form of design briefs, prepared from questions to the client and from discussion with the design team. Remember to ask as many questions as you possibly can to enable you to better understand what exactly the customer needs. Remember, a vague statement is equal to a vague understanding of the need, which may lead you to develop a product that does not match the customer needs. This is the most important step in developing the tree. Further discussion among the design team of what you would like to have in the product is important. Remember, there is no limitation for what you can put in the product at this time.

2. Order the list into sets of higher-level and lower-level objectives. The expanded list of objectives and subobjectives is grouped roughly into hierarchical levels.

3. Draw a diagrammatic tree of objectives, showing hierarchical relationships and interconnections. The branches in the tree represent relationships, which suggest means of achieving objectives.

3.3.1 Objective Tree: Water Purifier

After the residents of Gotham City complained to their mayor about the city's water quality, he ordered the health unit to investigate the complaint. The health unit's recommendation was based on the chemical analysis done by the top chemist at the Department of Public Health. The task is now given to the city engineers to design water purifiers. Your task is to help them build an objective tree.

Solution

1. Prepare a list of design objectives:
   a. Cost effectiveness
   b. Safety
   c. Can detect chemical imbalance
   d. Fewer repairs
   e. Easy to repair when needed
   f. Long lasting
   g. Affordable
   h. Low damage
   i. Low or no contamination
   j. Takes up least possible space
   k. Safe for humans
   l. Safe for environment
   m. Gets the job done

      n. Can correct problems in least time
      o. Low maintenance
      p. Cleans high volume of water
      q. Efficient
      r. . . . (complete per project objectives)
      s. . . .

2. Order the list into sets:

Safety	Cost Effectiveness	Efficiency
Safe for humans	Few repairs	Can detect chemicals
Safe for environment	Easy to repair	Long lasting
	Affordable	Low damage
	Takes least possible space	Gets job done
	Low maintainace	Corrects problems in minimal time

3. Draw an objective tree (Figure 3.1).

3.3.2 Objective Tree: Automatic Coffee Maker

Build an objective tree for an automatic coffee maker.

1. Prepare a list of design objectives:
      a. Safety
      b. Efficiency
      c. Quality
      d. Convenience
      e. Easy to use
      f. Fast
      g. Makes good coffee
      h. Doesn't burn user
      i. Good mixture
      j. Right temperature
      k. Splash proof
      l. Cheap to consumer
      m. Volume of coffee
      n. Automatic
      o. Timer
      p. Energy saver
      q. Temperature control
      r. Easy to clean
      s. . . . (complete per project objectives)
      t. . . .
      u. . . .

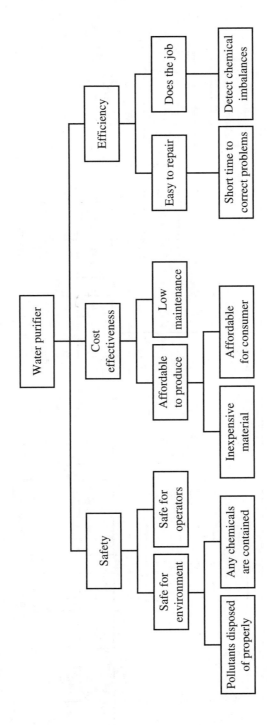

Figure 3-1　Objective tree for water purifier.

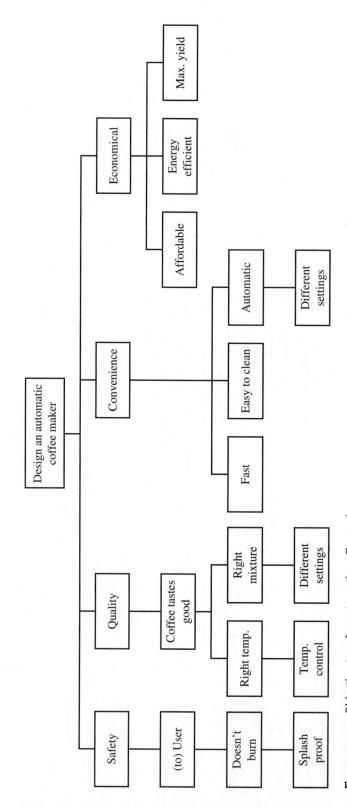

Figure 3.2 Objective tree for automatic coffee maker.

2. Order the list into sets:

Safety	Quality	Convenience	Economical
To user	Coffee tastes good	Easy to clean	Affordable
Doesn't burn user	Right temperature	Automatic	Volume of coffee
Splash proof	Good mixture	Timer	Energy saver
	Temperature control		Fast

3. Draw an objective tree. (See Figure 3.2.)

3.3.3   Objective Tree: Automatic Aluminum Can Crusher

We will use this example throughout the remaining chapters of this book. The example was offered as a design project. The project statement is as follows:

Design and build a device/machine that will crush aluminum cans. The device must be fully automatic (i.e., all the operator needs to do is load cans into the device; the device should switch on automatically). The device should crush the can automatically, eject the crushed can automatically, and switch off automatically (unless more cans are loaded). Students should adhere to the following guidelines:

- The device must have a continuous can-feeding mechanism.
- Cans should be in good condition when supplied to the device (i.e., not dented, pressed, or slightly twisted).
- The can must be crushed to one-fifth of its original volume.
- The maximum dimensions of the device are not to exceed $1.5 \times 1.5 \times 1$ feet.
- Performance will be based on the number of cans crushed in 1 minute.
- Elementary school children (K and up) must be able to operate the device safely.
- The device must be a stand-alone unit.
- The total cost of the device should not exceed the given budget ($200).

Although the need statement is relatively clear, the design teams did several interviews with the client, asked questions, and carefully listened to the client's responses in order to determine the goal of the intended device. Once the interview was completed, the engineering team met and constructed the objective tree shown in Figures 3.3 through 3.7, which clearly states the goals and defines the directions of the device design.

Figure 3.3   Main heading.

Figure 3.4   Performance branch.

Figure 3.5 Safety branch.

Figure 3.6 Appearance branch.

Figure 3.7 Inexpensive branch.

3.4 PROBLEMS

3.4.1 Team Activities

1. Use the statement "Connect two bodies of water with no gradient." Develop a series of question to better understand the need.
2. Use the following statement: "Ice is forming on the roof with density gradient." Develop a series of questions to better understand the need.
3. In this activity each group, as shown in the following list, asks as many questions as they can to clarify the given design statement:
   (a) Groups 1 and 9: Design a coffee maker.
   (b) Groups 2 and 10: Design a safe ladder.
   (c) Groups 3 and 11: Design a safe chair.
   (d) Groups 4 and 12: Design a safe lawn mower.
4. Draw an objective tree for an automatic tea maker.
5. Draw an objective tree for the following:

   If you drive in the state of Florida, you may notice some traffic congestion due to the trees being trimmed on the state roads and freeways. Assume that the Florida Department of Transportation is the client. Usually, branches 6 inches or less in diameter are trimmed. The allowable horizontal distance from the edge of the road to the branches is 6 ft. The material removed from the trees must be collected and removed from the roadside. To reduce the cost of trimming, a maximum of only two workers can be assigned for each machine. The overall cost, which includes equipment, labor, etc., needs to be reduced by at least 25% from present cost. The state claims that the demand for your machines will follow the price reduction (i.e., if you are able to reduce the cost by 40%, the demand will increase by 40%). Allowable working hours depend on the daylight and weather conditions.

6. Draw an objective tree for the following statement:

   We have a mountain-sized pile of wood chips that we want to process into home-use fire logs. We are located in Tallahassee, Florida. We can have a continuous supply of wood chips throughout the year. We should be able to produce 50 fire logs per minute. The shareholders require us to have a large profit margin, and our prices should be lower than our competition for the same size logs. The logs should produce enough heat to keep our customers faithful, and they should have environmentally clean exhaust.

7. As technology advances and new materials are synthesized, it seems that there are certain products that receive only minimal, if any, modernization attempts. Among these are the shopping carts that are used in grocery stores. There is a tendency to save parking spaces by not designating a return cart area. Leaving carts in the parking lots may lead to serious accidents and car damage. Furthermore, many customers do not fill their carts when shopping; however, they do not like to carry baskets. Other customers like to sort products as they shop. Develop an objective tree to clarify the need statement and prioritize the objectives laid out in the problem statement. You may add features that may not be listed clearly in the problem statement but will give an additional advantage to the proposed design.

8. Most houses have vents that open and close manually without any central control. Cities across the United States advise the use of such vents in an effort to save energy. In most cases, household occupants do not use the entire house at the same time; the tendency is to use certain rooms for long periods of time. For example, the family room and dining room may be used heavily, while the living room and kitchen are used at certain hours of the day. To cool or heat a room, the vent system must work to cool or heat the entire house. Energy saving can be enhanced if the vents of unused rooms are closed, this will push the hot/cold air to where it is needed most and reduce the load of the conditioning system. Develop an objective tree to clarify the need statement and prioritize the objectives laid out in the problem statement. You may add features that may not be listed clearly in the problem statement but will give an additional advantage to the proposed design.

3.5 SELECTED BIBLIOGRAPHY

1. Cross, N. *Engineering Design Methods: Strategies for Product Design*. New York: Wiley, 1994.
2. Dym, C. L. *Engineering Design: A Synthesis of Views*. Cambridge, UK: Cambridge University Press, 1994.
3. Hensel, E. "A Multi-Faceted Design Process for Multi-Disciplinary Capstone Design Projects," Proceedings of the 2001 American Society for Engineering Education Annual Conference and Exposition, Albuquerque, NM, 2001.
4. Karuppoor, S. S., Burger, C. P., and Chona, R. "A Way of Doing Design," Proceedings of the 2001 American Society for Engineering Education Annual Conference and Exposition, Albuquerque, NM, 2001.
5. Pahl, G., and Beitz, W. *Engineering Design: A Systematic Approach*. New York: Springer-Verlag, 1996.
6. Suh, N. P. *The Principles of Design*. New York: Oxford University Press, 1990.

Market Analysis and Information Gathering

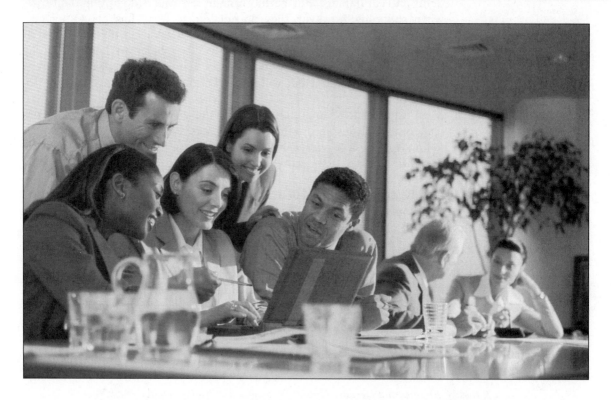

A market analysis is required at the start of the design process. Designers will use this market analysis to fine tune their designs. (Copyright Antonio Mo/Getty Images/PhotoDisc.)

4.1 OBJECTIVES

By the end of this chapter, you should be able to

1. Gather information terminology.
2. Identify SIC (Standard Industrial Classification) codes for the industry.
3. Identify data sources for your search.
4. Further clarify the statement of need.
5. Find what is available in the market that may provide you with leads to produce a better product.
6. Conduct market analysis for your product development.

In the design process, it is important to determine if a business opportunity exists for a new product before investing time and money to evolve the product. Most likely, products will fail when the market analysis only skims the surface. The analysis requires a thorough study of the total market. This includes its trend, competition, volume, profit, opportunities, consumer needs, and some indication of customer feeling for the product. Market analysis must be conducted at the start of the design process. It serves as a mechanism to define further the need statement and provides an opportunity for designers to review other attempts at solving the problem at hand, if they exist. Remember, the majority of designs are development designs. Thus, knowing what others have provided as a solution is very important before you attempt to offer your own solution.

It is also important at this stage not to provide solutions; you are gathering information with which to provide a better solution. This phase of the design process will help you appreciate your development and allow you to review what others have done in solving the same problem. It will also help you identify the size of the need and whom you are competing with. Further, it will show what is available on the market.

King defines a market as a group of potential customers who have something in common. Two different techniques are employed in market analysis:

1. *Direct search:* This involves obtaining information directly from the consumer, manufacturers, salespeople, and so forth. The information is collected by interviews and surveys.
2. *Indirect search:* Information is collected from public sources, such as patents, journal reports, government analysis, and newspapers.

The market search should be done in a systematic and objective manner and consider all information that is relevant to the product. It is important to remember that an unbiased outlook is necessary when analyzing the data to formulate a market analysis report. Any biases will be will be reflected in the report's view of the market, and the results could be disastrous.

4.2   HOW TO CONDUCT A MARKET ANALYSIS[1]

Thousands of resources, it seems, all present some variation of the same information. However, it is never presented in the form you need. Before beginning any research project, it is important to plan the process. Information collection is usually one of the first steps, and unfortunately it is often anticipated with the same enthusiasm as a root canal. For the engineering student, product/market research often requires the use of unfamiliar information sources and often produces huge quantities of information. It is important that engineering skills be applied when planning and dealing with this deluge. These skills include

- Critical thinking
- Strategy
- Analysis
- Time management (it always takes longer than you think)

In the product development process, market research is conducted initially to assess market potential, market segments, and product opportunities; and to provide production cost estimates and information on product cost, sales potential, industry trends, and customer needs and expectations. The search has the following steps:

1. **Problem definition**
   Knowing what to look for is very important before you start gathering information. This can be accomplished by answering questions similar to the following:
   a. What are the major objectives (review your objective tree) of the customer need?
   b. Are you developing a new product or solving a problem in an existing product? *Remember, you are not providing solution, you are redefining a problem.*
   c. In one sentence and in your own words (abstraction of the need statement), define the problem at hand.
   d. Who is your customer, and why would they want/need to buy the product (e.g., time saver, utility, unique value)?
   e. How are you going to go about getting your product to customers (e.g., development cost, time, manufacturing, production investment, etc.)?

2. **Develop a strategy**
   It is important to set a plan for the search process. Looking through every reference book in a library, merely hoping to find the right piece of information, will inevitably lead to an inefficient waste of time. It is helpful first to identify what pieces of information may be needed, and then to select where to begin the search. Searching can be accomplished by
   a. **Identifying keywords**
      It is important to have some relevant terms with which to begin the search.

---

[1] Information provided in this section is based on Suzanne Weiner's work at the Massachusetts Institute of Technology, Presentation at the ASEE Conference, Chicago, IL, 1997.

b. **Writing a plan**

It is important to realize that the search process is not linear. For example, while searching the business literature for information on industry trends, you may come across the text of an interview that directly identifies some customer needs for the product under investigation. This information can be very useful if properly contextualized; thus it is important to know the framework within which you are working.

3. **Information Checklist**

The following list can be used in two ways: as a planning tool for information collection, or as a checklist for locating relevant information.

a. Products
   i. Product names
   ii. Patents
   iii. Pricing
   iv. Parts breakdown
   v. Product features
   vi. Development time

b. Companies
   i. Major players
   ii. Company financials for major players
      • Annual reports
      • 10K reports

c. Industry
   i. Trends
   ii. Labor costs
   iii. Market size
   iv. Industry facts: pieces of information gained from various sources that help to clarify anything about the industry

d. Market information
   i. Market reports
   ii. Market share of major companies in industry
   iii. Target markets of major competitors
   iv. Demographics:
      • Age
      • Geographic location
      • Gender
      • Political/social/cultural factors

e. Consumer trends

4.3   RELEVANT INFORMATION RESOURCES

There are literally hundreds of resources in which information may be located, but several stand out due to their superiority of scope, quality, and overall usefulness.

The resources can be divided into three categories:

1. Industry information
2. Company information
3. Market information

Most journal databases will provide information relevant to all of the areas listed in previous section, and this should be kept in mind when searching and analyzing. This will also account for the repetition of resources in all three areas.

4.3.1 Industry Information

It is essential to know information about the industry your company is considering entering. Who are the major players? What are the current trends? How large (in dollar amounts) are the relevant industries? What materials are used by the relevant industries? The following resources can provide access to information that will answer these questions:

1. **Standard industrial classification (SIC) code**
   Many business information sources are organized by the SIC codes. The following publications are guides to these codes:
   a. *Standard Industrial Classification Manual.* The official U.S. government manual that provides SIC codes at the 2-digit and 4-digit levels.
   b. Web version of the SIC code. This provides access only by SIC category, not subject.

2. **Trade associations**
   a. *Encyclopedia of Associations.* Gives address, telephone, and contact information as well as a brief description of trade and industry associations and their publications. Descriptions usually include dates/places for conferences and trade shows.

3. **Industry overview**
   a. *Manufacturing U.S.A.: Industry Analyses, Statistics, and Leading Companies.* This annual publication provides statistical data on 459 manufacturing industries, including ratios and occupational data. State-level data are included.
   b. *Forbes.* Provides an annual report on American industry in the January issue of each year.
   c. Standard and Poor's industry surveys. Provide a textual analysis for 22 broad industry categories, which are broken down into more specific subsets.
   d. *Encyclopedia of American Industries.* This publication is organized by SIC code with a list of related publications for each described industry.
   e. *Moody's Industry Review.* This statistical source contains key financial information and operating data on about 3500 companies. The information is arranged by industry, in 137 industry groups. The companies within an industry group may be compared with one another, and they may also be measured against certain averages for that industry. The statistics are updated twice each year.
   f. *Predicates Basebook.* Dewey Library holding began with 1977, but in the 1983 volume, the statistics go back to 1967, and they have been compiled annually since 1970. This annual statistical source contains about 29,000 time series, which are arranged by a proprietary 7-digit SIC code–based system. The data include economic indicators and industry statistics. The industry statistics usually include production, consumption, exports/imports, wholesale price, plant and equipment expenditures, and wage rate.

g. *Predicasts Forecasts.* This quarterly statistical source provides short- and long-range forecasts for economic indicators as well as industries and products. The data are arranged by a proprietary 7-digit SIC code–based system. Each forecast includes the date and page reference of the source from which the data are taken. The front of each annual accumulation includes composite forecasts, which present historical data for over 500 key series.

4. **Statistics**
   a. Annual Survey of Manufacturers. This survey supplements the Census of Manufacturers in USA.
   b. American Statistics Index (ASI). Indexes and abstracts statistics published by the U.S. government.
   c. Statistical Reference Index (SRI). Indexes and abstracts statistics published by trade and industry organizations and state governments.

4.3.2   Company Information

Company sources provide information on the players within an industry. They provide information on the company financials, products (primary and secondary such as Pepsi and Mountain Dew), brands/trade names, and other "nuggets" that, in context, can help with overall market assessment and analysis.

When searching for company information, it is important to determine whether the company is public or private. Public companies are required to register and file reports with the Securities and Exchange Commission (SEC), while private companies are not. Generally, it is much easier to find information on public companies.

Some companies publish their 10K information on the Web. Examples include

1. Allied Products Corporation
2. American United Global, Inc.

The Web pages contain information extracted from annual and 10K reports of over 12,000 public companies. The data are compiled into a standard format, which allows for sophisticated manipulation of data and customized display formats.

Directories

1. *Hoovers Corporate Directory.* Provides the company's name, street address, phone number, and fax number; the names of the Chief Executive Officer, Chief Financial Officer, and Human Resources director; the most recently available annual sales figures; the percentage change in sales from the previous year; the number of employees; a description of what the company does; and the company's status (private or public).
2. *Directory of Corporate Affiliations.* Lists over 4000 parent companies with divisions, subsidiaries, and affiliates. Geographic and SIC indexes.
3. *Duns Million Dollar Disc.* Lists over 160,000 U.S. businesses with net worth over $500,000. Can be used to identify companies by state, industry, or geographic location. Provides thumbnail bios of top executives for public and private companies and basic company information.

4. *Standard & Poor's Register of Corporations, Directors, and Executives, Volume 1; Corporations.* Corporate addresses, telephone numbers, names and titles of officers, SIC codes, products, annual sales, and number of employees for 55,000 public and private companies. Volume 3 provides an index by SIC code.

5. *Corporate Technology Directory.* Directory for high-tech companies with very specific product indexes. Provides basic company information.

6. *Thomas Register of American Manufacturers and Thomas Register Catalog File.* Volumes 1 through 8 list companies arranged by products and services. Volumes 9 and 10 provide an alphabetic list of manufacturers with addresses, subsidiaries, products, and asset estimates; Volume 10 provides an index of brand names. Selected company catalogs are reproduced in Volumes 11 through 16.

7. *Verizon Yellow Pages.* This free service that allows users to look up names, addresses and phone numbers of over 2.1 million businesses.

8. *Ward's Business Directory.* Ward's is a good source for private companies. It provides information on 142,000 companies, 90% of which are private. Volumes 1 through 4 provide individual company information. Volume 5 ranks companies by SIC code.

Corporate Reports. Moody's manuals also provide detailed information on company history, subsidiaries, business and products, comparative balance sheets, stock and bond descriptions, and more. Geographic indexes classify companies by industry and product. The data in Moody's manuals are obtained from annual and 10K reports.

4.3.3 Market Information

Market information may be found in many places. The search for information should make use of the related industry, company, and product keywords. This is true when using either a paper source or an electronic source. Some buzzwords include market share, target market, demographics, and market potential.

Market research reports are ready-made reports that have been commissioned by companies or industrial groups. They are compiled by one of several market research companies.

Market Research Reports. These are reports provided by consulting & advisory firms on companies and different sectors of industry. The information can be found from firms such as:

1. Findex
2. Frost & Sullivan Reports

Market Share and Other Information

1. *Market Share Reporter.* Annual publication providing market share data for companies, products, and industries.
2. *World Market Share Reporter.* Annual publication providing market share information.
3. *Statistical Abstract of the United States.* Statistics on the industrial, social, political, and economic aspects of the United States. A great source for all types of useful statistical information.

4. *Industrial Statistics Yearbook.* Industrial statistics published by the Department of Economic and Social Affairs, Statistical Office of the United Nations.

Demographic Information

1. *American Demographics.* This journal is searchable on ABI/Inform on First-Search.
2. *State and Metropolitan Area Data Book.* Provides social and economic data for SMSAs, cities, states, and census regions. Prepared by the U.S. Bureau of Census.
3. *Demographics USA, County Edition.* (http://www.tradedimensions.com) Contains information on purchasing power, consumption rates, economic conditions, and population statistics.

4.4 WEB TOOLS

At this time the Web, by itself, does not provide access to all of the resources needed to conduct a product/market research. However, there are some great sites that point you to the resources that are available:

1. Business Researcher's Jumpstation at http://www.brint.com/sites.htm
2. Marketing Resource Directory at http://www.ama.org
3. MIT Resource List at http://me.mit.edu/resources/

Web search engines are increasing in sophistication, as are the Web sites of companies, distributors, industrial associations, and retailers. Therefore, it is valuable to spend some time sifting through the information available on the Web.

4.5 MARKET ANALYSIS—SAMPLE

Consider the example of the automatic can crusher. Here is a short summary of the market analysis.

4.5.1 Problem Definition

The client requested a device that is able to crush aluminum cans. The device must be fully automatic. All the operator will need to do is load the cans into the device. It will then switch on automatically, crush the cans, and switch off automatically. The device must have a continuous can-feeding mechanism. Cans must be crushed to one-fifth of their original volume. The maximum dimensions of the device must not exceed $1.5 \times 1.5 \times 1$ feet. Performance will be based on the number of cans crushed in one minute. Elementary school children must be able to operate the device safely. The device must be a stand-alone unit.

4.5.2 Potential Customers

- Schools
- Colleges
- Hospitals

- Hotels
- Resorts
- Shopping malls
- Playgrounds and recreational areas
- Apartments, dormitories
- Sports arenas
- Office buildings
- Residential homes

4.5.3   Companies

- Edlund Company, Inc. (159 Industrial Parkway, Burlington, Vermont). Mr. R. M. Olson (President)
- Prodeva Inc. (http://prodeva.com)
- Enviro-Care Kruncher Corporation (685 Rupert St., Waterloo, Ontario N2V1N7, Canada)
- Recycling Equipment Manufacturer (6512 Napa, Spokane, Washington 99207)
- Kelly Duplex (415 Sliger St., P.O. Box 1266, Springfield, Ohio 45501)
- Waring Commercial (283 Main St., New Hartford, Connecticut 06057)
- DLS Enterprises (P.O. Box 1382, Alta Loma, California 91701)

4.5.4   SIC Code

- Food service industry 3556
- Recycling 3599

4.5.5   Trends

The aluminum industry produces 100 billion cans a year. This number has been flat for the past three years. In 1997, 66.8 billion of the cans were returned for recycling. This figure showed an increase of about 6.3% over the 1996 figure. The industry collected 2.05 billion pounds of aluminum in 1997.

4.5.6   Market Information

Growth opportunities for aluminum beverage cans exist throughout the world. Global aluminum can shipments are forecasted to increase 5% annually to 209 billion by the year 2000. Seventy-four percent of U.S. beverage sales occurred in convenience stores, drugstores, clubs, mass merchandise stores, vending machines, and grocery markets. For this product, the initial market is schools. Leon County, Florida public schools, for example, at the high school level, have around 48,846 students.

4.5.7   Patents

Around 56 patents are listed for the past 10 years. A Web search using the IBM patent search engine will show the details.

4.6 PROBLEMS

4.6.1 Team Activities

1. How is the direct search different from the indirect search?
2. Define market analysis and discuss how is it different from information gathering.
3. Develop a strategy with which your team will conduct the market analysis task.
4. Write a plan to conduct your market analysis.
5. Estimate the time required for your team to conduct the market analysis assignment.
6. Consider the following need statement:

> Most houses have vents that open and close manually without any central control. Cities across the country advise the use of vents to help save energy. In most cases, household occupants do not use the entire house at the same time; the tendency is to use certain rooms for a long period of time. For example, the family room and dining room may be used heavily, while the living room and kitchen are used at certain hours of the day. To cool or heat a room, the system must to work to cool or heat the entire house. Energy saving can be enhanced if the vents of unused rooms are closed; this will push the hot/cold air to where it is needed most and reduce the load of the air conditioning system.

A design team has performed market analysis for this statement but forgot to fill in all of the information. Gather information from available sources and complete the following market analysis:

I. The SIC codes that are associated with this design are
   a. Airflow controllers, air conditioning and refrigeration valve manufacturing (provide the codes)
   b. Air conditioning units; domestic and industrial manufacturing (provide the codes)
   c. Air ducts, sheet metal 5039
II. Materials used in the industry are as follows:
   a. Sheet metal
   b. (provide the codes)
   c. (provide the codes)
   d. (provide the codes)

 III. Major associations that deal with air conditioning are as follows:
  a. Air Conditioning and Refrigeration Institute
  b. (provide the codes)
  c. (provide the codes)
  d. (provide the codes)
 IV. Major companies
  a. Trane
  b. (provide the codes)
  c. (provide the codes)
 V. Patent search
  a. Sarazen et al., 4,493,456
  b. (provide the codes)
  c. (provide the codes)

4.6.2 Individual Activities

1. Perform a Web search and name three industries that change a product at a frequency of at least once each year.
2. How many patents were filled in the last 10 years on money sorters? List a few and identify their differences.
3. List as many Web addresses as you can find that provide patent search. Name some differences among the sites.
4. Interview a librarian in your school and identify research sources that are available for your use.

4.7 SELECTED BIBLIOGRAPHY

1. http://www.osha.gov/
2. http://www.ntis.gov/
3. http://www.census.gov/epcd/www/naicstab.htm
4. http://vancouver-webpages.com/global-sic/
5. http://www.thomasregister.com/
6. http://www.tipcoeurope.com/
7. http://iml.umkc.edu/ndt/html/lexus.html
8. http://www.lexis-nexis.com/lncc/
9. http://www.google.com
10. http://www.yahoo.com
11. http://www.briefing.com/
12. http://www.sme.org/
13. http://www.asme.org
14. http://www.manufacturing.net/
15. http://stats.bls.gov/blshome.htm
16. http://www.uspto.gov/
17. http://www.european-patent-office.org/
18. http://www.nist.gov/
19. http://www.entrepreneur.com/

20. http://www.tenonline.org/
21. http://www.tradedimensions.com
22. http://www.findex.com
23. King, W. J. *The Unwritten Laws of Engineering—Mechanical Engineering*, Vol. 66, No. 7, 1944.

Establishing Functional Structure

It is important to determine a product's functional structure before completing the design. Engineers will determine the function of each of this drill's components in order to determine all possible solutions in its final design. (Copyright Craig Brewer/Getty Images/PhotoDisc.)

5.1 OBJECTIVES

By the end of this chapter, you should be able to

1. Discuss the essentials of function analysis.
2. Create function structures based on need statements.
3. Discuss the importance of function analysis.

At present there are several design strategies that are utilized by design teams. In many cases engineering designers develop their own methods, usually based on their experience and current practices in their working environment. However, when engineers are faced with a novel problem outside their immediate experience, they search furiously for an organized systematic procedure. Several theories and methodologies have been proposed for design. In one theory it is argued that designers are born with a creative mind that is capable of producing inspired designs. In this theory, design cannot be taught or learned; it is a gift bestowed on unique and privileged individuals. Fortunately, this theory is not widely accepted. In another design philosophy, once a need is identified a concept is directly developed by brainstorming. However, this philosophy often does not lead to quality designs, and it does not foster the use of the latest technologies. In this methodology designers have a tendency to take their first idea and keep refining it toward a product design. In this book we utilize a more scientific view of design, a systematic approach in logical sequence. This approach is in agreement with what is known as the "industry best practice".

In the process you have followed thus far, you are still defining the problem utilizing different tools. Thus far you have developed an objective tree; you have searched the market; and you need to start developing an idea on how the product could function. Now you need to generate a functional decomposition tree from the problem statement based on the objective tree.

5.2 FUNCTIONAL DECOMPOSITION

Functional decomposition supports the divergent–convergent design philosophy concept. This concept requires that a design problem be expanded into many solutions before it is narrowed to one final solution. All design problems should be decomposed into the design of functionally independent subsystems. In this chapter, we develop techniques for decomposing any problem into smaller, more easily manageable parts by understanding the functions required of the device and treating each function as a separate subsystem.

Solution principles or designs based on conventional methods do not provide optimum designs when new technologies and discoveries are to be used. The crux of a problem should be expressed as a list of the required functions and essential constraints. The first step in function decomposition is to determine the overall function, subfunction, and sub-subfunction of the design problem.

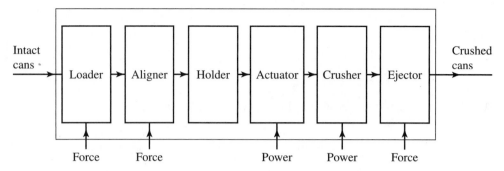

Figure 5.1   Function tree for the automatic can crusher.

Let's refer to our example of an automatic can crusher. The design team reviewed the objective tree and their market analysis findings before they started structuring a function tree. They kept their minds open to all possible methods for accomplishing the stated goals. The function tree is shown in the Figure 5.1.

As you observe from the figure, the system is enclosed with a boundary that allows a flow of energy, input of cans, and output of crushed cans to enter through the boundary. There are a series of functions connected from the input side to the output side that compose the whole system. A loader is utilized to input the cans into the system. The aligner will position the cans inside the device for crushing. A holder will fix the can in place during the crushing. An actuator will send a signal to the crushing device. Then, after the crusher, an ejector must remove the cans from the system.

You should ask at this point, "Is this the only set of functions that could perform the essentials of crushing cans?" The answer is no. However, it should be emphasized that the systematic design process is an iterative one. Further proposing a system to crush cans in terms of functional needs will satisfy the customer need statement, which is redefined through the objective tree. There could be some functions and requirements in the objective tree that no action function could satisfy, such as the objective of having a safe device. This requirement, however, will be further clarified in later functions. At this stage the design team should revisit the objective tree and revise the tree if additional functions are found to be necessary. Thus the iterative process is demonstrated.

Design teams should be alerted that they do not generate an idea and then fit the function analysis to that one idea. At this stage you should have no constraints on what could be implemented in the design and what could not. Students who tend to bend the function analysis to fit an agreed-on and hidden idea tend to spend more time refining the one idea to fit the procedures and steps needed in the future stages of the design process. Creating function is an essential step in the design process; the later design steps depend on the function analysis step. Thus revisiting this step and reiterating is highly recommended.

5.3 FUNCTION STRUCTURE

5.3.1 Overall Function Diagram

The overall function diagram is presented by a black box that shows the inputs and outputs to the system, including the flow of energy and material and information from/to the system surroundings. The functions and subfunctions will not appear in this diagram. This form of presentation for the function tree expresses the relationship between inputs and outputs regardless of the solution within the box. Figure 5.2 shows the overall function diagram for a deformation testing machine.

5.3.2 Function Blocks

A complex, overall function describing the problem statement can be broken into a number of functions and subfunctions. Many of the subfunctions can be divided into sub-subfunctions. The division depends on the type of design system being considered. In original designs, neither the individual subfunctions nor their relationships are clearly known. Hence, in such a case, a strong effort to search for and establish an optimum function structure is required. However, in adaptive designs, general function structure and the functional relationships between various components and systems are usually well established.

The lower levels of functions carry smaller levels of complexity and should be simple to design. Function structures are of great importance in the development of modular designs. An important advantage of a function structure is that each of the components and systems can be examined individually and modified if needed. Function structures are very useful when designing and using design catalogs because for every function, a task-specific component or system can be identified.

An example of function structure for a grocery cart generated using an objective tree (see Chapter 3) is shown in Figure 5.3 (on the next page).

Figure 5.2 Overall function.

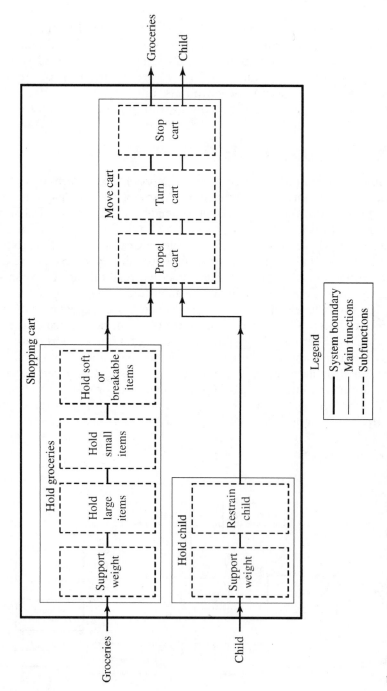

Figure 5.3   Function structure for grocery cart.

As shown in Figure 5.3, there are three major functions, and each has several subfunctions. They can be summarized as follows:

1. Hold groceries
   a. Support weight
   b. Hold large items
   c. Hold small items
   d. Hold soft or beverage items
2. Hold child
   a. Support weight
   b. restrain child
3. Move cart
   a. Propel cart
   b. Turn cart
   c. Stop cart

The system inputs are groceries and child, and the outputs are the same. The overall function is surrounded by a boundary that is presented by a solid line; each of the functions is also blocked with a solid line, and the subfunctions are surrounded by dashed lines. This will keep the boundaries well defined for all functions and subfunctions.

Now let us look at an example of determining the functional structure of an existing design. Consider a paper stapler. Students are encouraged to take apart an existing paper stapler and identify the different components. Then identify the specific task of each of the components. Either individually or collectively, the components serve a specific function or subfunction, as seen in the Figure 5.4. Match your components to the listed functions.

Another example is a Black & Decker drill. A sketch of a disassembled drill is shown in Figures 5.5 and 5.6. Students are asked to name the functions of the drill components.

Some of the basic points in generating functional diagrams are as follows:

1. From the objective tree, establish the main function, a few subfunctions, and some sub-subfunctions for each of the subfunctions. A step-by-step break down at each level is needed.
2. Arrange the subfunctions so that there is some relationship between them. Do not worry about the physical or logical relationship at this stage.
3. Determine the flow of energy, materials, and information. Find the main and auxiliary flow of these three items.
4. Establish a structure such that the main function is on top of the tree, branches into subfunctions, and then repeats to sub-subfunctions.
5. Function structures should be kept as simple as possible. Simplicity leads to simple and economical solutions.

The primary purpose of function structures is to facilitate the discovery of solutions. Although the structure should not presuppose any solution, it is legitimate to conceive of a preliminary solution and abstract the solution. However, through a process of iteration, the structure should be made generic to prevent precluding new and innovative solutions.

Task-specific functions, provided in the following list, can be utilized in generating a functional analysis:

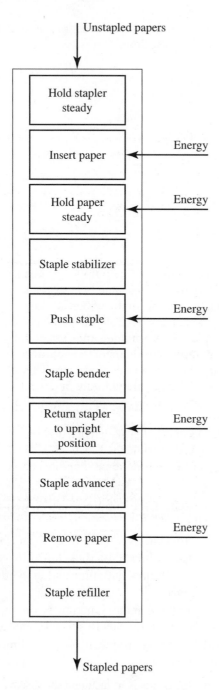

Figure 5.4 Analysis of paper stapler function structure.

Figure 5.5   Schematic of Black & Decker drill.

1. Conversion of energy:
   a. Changing energy (e.g., electrical to mechanical energy)
   b. Varying energy components (e.g., increasing speed, torque)
   c. Storing energy (e.g., storing potential or kinetic energy)
   d. Connecting energy with information (e.g., switch to start a motor)
2. Conversion of materials
   a. Changing matter (e.g., melting [solid to liquid])
   b. Connecting matter with energy (e.g., moving parts)
   c. Rearranging materials (e.g., mixing or separation)
   d. Storing materials (e.g., storing material in a silo)
3. Conversion of information
   a. Changing signals (e.g., mechanical to electrical)
   b. Connecting information with energy (e.g., amplifying signals)
   c. Connecting information with matter (e.g., marking metals)
   d. Storing signals (e.g., data banks)

Figure 5.6 Schematic of Black & Decker drill.

5.4 PROCEDURE TO ESTABLISH FUNCTIONAL STRUCTURES

1. *Express the overall function for the design in terms of conversion of inputs and outputs.* The key is to determine what needs to be achieved by the new design, and how it is to be achieved. This can be accomplished by representing the product or device (to be designed) simply as a black box, (Figure 5.7), which converts given inputs to desired outputs. The black box specifies all the functions required to convert the input into output. It is important not to place any restrictions on the function. Such restrictions limit the solution space or system boundary. For example, the function "push from A to B" restricts the solutions from ejecting, throwing, rolling, lifting, or even catapulting the object from A to B. Thus, it is important to widen the solution space or system boundary as much as possible.

   It is important to try to ensure that all the relevant inputs and outputs are listed clearly. As mentioned earlier, they can be classified as flows of materials, energy, and information.

Inputs   Functions   Outputs

The black box systems model

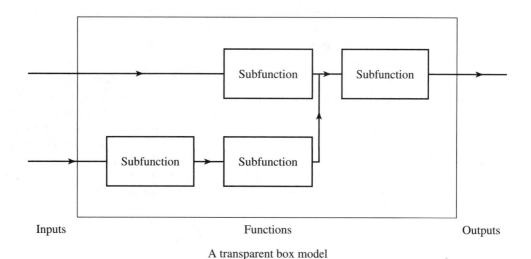

Inputs   Functions   Outputs

A transparent box model

Figure 5.7   The box diagram.

2. *Break down the overall function into a set of essential subfunctions. (Figure 5.7)*
   The conversion of the set of inputs into the set of outputs at the main function
   level is usually very complex. Thus the black box needs to be broken into subtasks or subfunctions. The breakup depends on the experience of the designer,
   availability of components capable of performing specific tasks, and the latest
   technologies. In specifying subfunctions, it is necessary to express the task as a
   verb plus a noun (e.g., amplify signal, count items, decrease speed, increase diameter or an action noun such actuator, loader, etc.). Each subfunction has its
   own input(s) and output(s).

3. *Draw a block diagram showing the interactions between subfunctions.* A
   block diagram consists of all subfunctions separately identified by enclosing
   them in boxes. They are linked by their inputs and outputs in order to satisfy
   the overall function of the product or device that is being designed. In other
   words, the original black box of the overall function is redrawn as a transparent box in which the necessary subfunctions and their links can be seen.

   When drawing this diagram, we decide how the internal inputs and outputs of the subfunctions are linked so that we may make a feasible working
   system. We may find that inputs and outputs may have to be juggled and
   perhaps redefined, and some subfunctions may need to be reconnected. It is
   advised that different types of lines (continuous, dotted, and dashed lines)
   be drawn to indicate the different types of flows (material, energy, and information) within the block diagram.

4. *Draw the system boundary.* In drawing the block diagram, we also need to make decisions about the precise extent and location of the system boundary. For example, there can be no loose inputs or outputs in the diagram, except those that cross the system boundary.

The boundary now needs to be narrowed again because of its earlier broadening during the consideration of inputs, outputs, and overall function. The boundary has to be drawn around a subset of the functions that have been identified in order to define a feasible product. It is also possible that the designer may not have complete freedom when drawing the system, since it may be restricted by the customer.

5. *Search for appropriate components for performing each subfunction and its interactions.* If the subfunctions have been defined adequately and at an appropriate level, then it should be possible to identify a suitable component for each subfunction. The identification of a component will depend on the nature of the product or device. For example, a component can be a mechanical, electrical, or combination device. A microprocessor-based electronic device can substitute some older versions of electromechanical devices. Since the function analysis method focuses on functions, new devices can be substituted at later stages of design or development.

5.4.1   Example of a Washing Machine (Adapted from Pahl & Beitz)

A simple example of the use of function analysis method is the design of the domestic washing machine. The overall function of the machine is to convert an input of soiled clothes into an output of clean clothes, as shown in Figure 5.8 (on the next page).

Inside the black box a process that separates the dirt from the clothes must occur, and the dirt itself must also be a separate output. We know that the conventional process involves water as a means of achieving this separation. Thus, a further stage must be added to convert clean but wet clothes to clean and dry clothes. Even further stages involve pressing and sorting clothes.

The essential subfunctions, together with the conventional means of achieving them (for converting soiled clothes into clean, pressed clothes), would therefore be as follows:

Essential Subfunctions	Means of Achieving Subfunctions
Loosen dirt	Add water and detergent
Separate dirt from clothes	Agitate
Remove dirt	Rinse
Remove water	Spin
Dry clothes	Blow with hot air
Smooth clothes	Press

The development of the washing machine involves progressively widening the system boundary, as shown in Figure 5.8. Early washing machines simply separated dirt from clothes but did nothing to remove excess water from the clothes. The

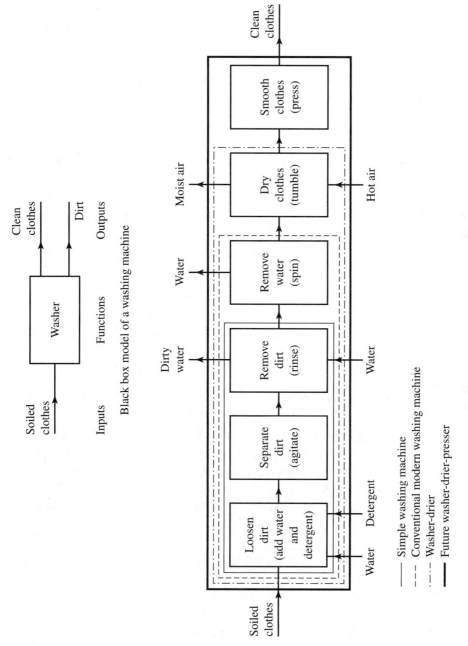

Figure 5.8 Function analysis of washing machine.

process of wringing the clothes dry was left to the human operator. The inclusion of a spin-drying function removed excess water but still left a drying process. This process is now incorporated into a washer-dryer system. Perhaps the smoothing and ironing of clothes should be incorporated in future designs, in addition to the current process.

5.5   FUNCTIONS

5.5.1   Guidelines on Identifying Functions

Functions should consider what and not how, and they should break the functions down as finely as possible. Functional decomposition is represented as follows:

A box
An action verb
An object on which the action verb takes place (verb–noun combination)
Known flow of materials, energy, and information

5.5.2   Essential Features

List all the alternative functions.
Include all the inputs and outputs.
For redesign problems, disassemble the sample to find functions.

5.6   SUBFUNCTIONS

The flow should be in a logical or temporal order.
Identify redundant functions and combine them.
Functions not within the system should be eliminated.

Specifically, there are flows associated with materials, energy, and information. Some functions associated with the flow of materials are as follows:

1. Material can be designed to alter its position or shape. Usual action verbs are *lift*, *position*, *hold*, *support*, *move*, *translate*, *rotate*, and *guide*.
2. Material can be divided into two or more bodies. The action verbs are *disassemble* and *separate*.
3. Material assembly. Usual terms are *mix*, *attach*, and *position relative to*.

Functions associated with the flow of information are usually in the form of mechanical or electrical signals, or software.

Functions concerned with the flow of energy in electromagnetic systems are mechanical, electrical, fluid, and thermal. Energy of this type of can be supplied, stored, transformed, or dissipated.

5.7   FUNCTIONAL STRUCTURE—SAMPLE

A design team was asked to develop a newspaper vending unit that is able to dispense three types of papers and not be easily vandalized. The design team proposed the following functional structure for the unit. The unit input is papers and the output is sold paper. The following are the functions and subfunctions as suggested by the design team:

1. Load the papers
2. Align the papers
3. Hold the papers
4. Payment
    a. Payment acceptor
    b. Payment counter
    c. Payment trigger
5. Release a paper
6. Dispense a paper

Figure 5.9 shows the function structure for the system.

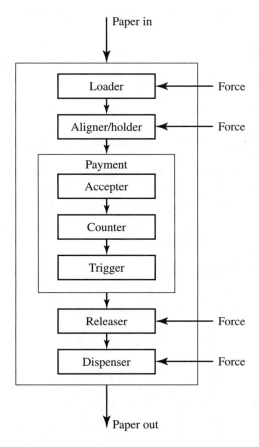

Figure 5.9 Function tree for the newspaper vending machine.

5.8 PROBLEMS

5.8.1 Team Activities

1. Discuss the following statement as it relates to the stage of the design process you have achieved thus far: The design process is divergent–convergent process.
2. As was discussed in this chapter, there are several trends and philosophies on design. Perform a technical search and list four different trends. Discuss the differences and similarities among these trends.
3. Why would it be wrong to state a mechanism of performing a function rather than generalize and name the function in building up a function structure?
4. What is the difference between functions and subfunctions?
5. Would it be possible for different design teams to come up with different function structures for the same system/product? Why or why not?
6. As a team, choose one of the following products to take apart. Name the parts and what function they serve. Then build a functional structure for the product.
   a. Paper hole puncher
   b. Cellular phone
   c. Computer mouse
   d. Umbrella
   e. Tape dispenser
   f. Can opener
7. Develop a function tree to produce a device that is capable of converting wood chips to fire logs.
   a. Express the overall function for the system.
   b. Break down the over all function into a set of subfunctions.
   c. Draw a block diagram showing the interactions between the subfunctions.
   d. Draw the system boundary.

5.8.2 Individual Activities

1. List the steps needed to generate functional structure.
2. What is the difference between functional structure and an objective tree?
3. Why can't you find a function for each objective you generate in the objective tree?
4. What function would you propose to satisfy the "must be safe" objective for the newspaper vending machine?
5. Generate a function tree for a coin sorter device.
6. Generate a function tree for a leaf-removing device.

5.9 SELECTED BIBLIOGRAPHY

1. Birmingham, R., Cleland, G., Driver, R., and Maffin, D. *Understanding Engineering Design.* Upper Saddle River, NJ: Prentice Hall, 1997.
2. Burghardt, M. D. *Introduction to Engineering Design and Problem Solving.* New York: McGraw-Hill, 1999.
3. Cross, N. *Engineering Design Methods.* New York: Wiley, 1994.
4. Cross, N., Christian, H., and Dorst, K. *Analysing Design Activity.* New York: Wiley, 1996.
5. Dym, C. L. *Engineering Design: A Synthesis of Views.* Cambridge, UK: Cambridge University Press, 1994.
6. Eder, W. E. "Problem Solving Is Necessary, But Not Sufficient," American Society for Engineering Education (ASEE) Annual Conference, Session 2330, Milwaukee, WI, 1997.
7. Eder, W. E. "Teaching about Methods: Coordinating Theory-Based Explanation with Practice," Proceedings of the American Society for Engineering Education (ASEE) Conference, Session 3230, Washington, DC, 1996.
8. Eekels, J., and Rooznburg, N. F. M. "A Methodological Comparison of Structures of Scientific Research and Engineering Design: Their Similarities and Differences," *Design Studies*, Vol. 12, No. 4, pp. 197–203, 1991.
9. Fulcher, A. J., and Hills, P. "Towards a Strategic Framework for Design Research," *Journal of Engineering Design*, Vol. 7, No. 2, pp. 183–194, 1996.
10. Jansson, D. G., Condoor, S. S., and Brock, H. R. "Cognition in Design: Viewing the Hidden Side of the Design Process," *Environment and Planning B, Planning and Design*, Vol. 19, pp. 257–271, 1993.
11. Holt, K. "Brainstorming from Classics to Electronics," Proceedings of Workshop, EDC Engineering Design and Creativity, edited by W. E. Eder, Zurich, pp. 113–118, 1996.
12. Hubka, V., Andreasen, M. M., and Eder, W. E. *Practical Studies in Systematic Design.* London: Butterworths, 1988.
13. Hubka, V., and Eder, W. E. *Theory of Technical Systems.* New York: Springer-Verlag, 1988.
14. Hubka, V., and Eder, W. E. *Design Science: Introduction to the Needs, Scope and Organization of Engineering Design Knowledge.* New York: Springer-Verlag, 1996.
15. Kuhn, T. S. *The Essential Tension: Selected Studies in Scientific Tradition and Change.* Chicago: University of Chicago Press, 1977.
16. Pahl, G., and Beitz, W. *Engineering Design: A Systematic Approach.* New York: Springer-Verlag, 1996.
17. Schon, D. A. *The Reflective Practitioner: How Professionals Think in Action.* New York: Basic Books, 1983.
18. Suh, N. P. *Principles of Design.* Oxford University Press, 1989.
19. Tuomaala, J. "Creative Engineering Design," Proceedings of Workshop, EDC Engineering Design and Creativity, edited by W. E. Eder, Zurich, pp. 23–33, 1996.
20. Ulman, D. G. *The Mechanical Design Process.* New York: McGraw-Hill, 1992.
21. Vidosic, J. P. *Elements of Engineering Design.* New York: The Ronald Press Co., 1969.
22. Walton, J. *Engineering Design: From Art to Practice.* New York: West Publishing Company, 1991.

6 Specifications

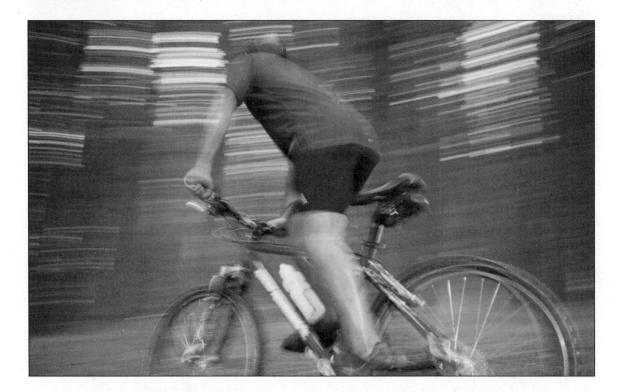

At the specification stage of the design process, the engineer can continue to provide clarification of the needs statement. This rider needs to stay dry when biking through a wet forest. How can the bike's splashguard be designed to fill such a need? (Copyright Ryan McVay/Getty Images/PhotoDisc.)

6.1 OBJECTIVES

By the end of this chapter, you should be able to

1. Quantify qualitative objectives.
2. Utilize techniques to organize specifications into categories.
3. Further clarify the need statement.

At specification stage of the design process you continue to provide additional clarification of the need statement. The objective tree provided a tool with which to list customer needs and design team objectives into categories. The function analysis provided a mechanism with which to activate the objectives without setting limits or quantifying objectives. Specifications are important to a successful design and may be considered the last step in defining the problem. The objective tree defines the vague statements that are introduced by the customer. But the objective tree and the structural functions do not set specific limits on the different functions and objectives. The objective tree or function structures are statements of what a design must achieve or do, but they are not normally set in terms of precise limits, which is what a performance specification does. For example, an objective may be stated to develop a product, which has low weight. Yet the term *low weight* is not clearly defined. Does it mean 20 lb or 100 lb?

Some customers (10 to 20% of customers, on average) are able to provide specifications that set limits and clarify vague requirements. An example of such is an announcement from the National Inventors Council for the design of a stair-climbing wheelchair:

NATIONAL INVENTORS COUNCIL
U.S. DEPARTMENT OF COMMERCE
WASHINGTON, D.C. 20550

There is a great need for a STAIR-CLIMBING WHEELCHAIR. The objective of such a chair is to give an active handicapped individual an additional range of mobility. The desired design should enable the individual to cope with the usual problems he or she might encounter when traveling to and from work, moving about industrial buildings, and the like. It must be remembered that the chair will perform the usual wheel chair function approximately 95% of the time; therefore, not too much of the conventional wheelchair's versatility and convenience should be sacrificed in providing the climbing function. In this connection it might be well to point that many active handicapped are able to fold a conventional wheelchair, put it in an automobile and drive to work. If this cannot be accomplished with a climbing chair, the overall objectives of providing the handicapped with independent mobility will not be achieved.

The following factors are to be taken into consideration:

(a) WEIGHT OF OCCUPANT: Assumed to be 200 pounds.
(b) WEIGHT OF CHAIR: 50 to 75 pounds, maximum.

(c) COLLAPSIBILITY: Capable of being folded by user and stowed in the interior of a standard automobile or taxi cab. Special loading ramps, which are not carried, as a part of the chair system, or special modifications to the automobile to accommodate chairs will not be considered.

(d) WIDTH OF CHAIR: 25 inches maximum. The American Standards Association utilized this width as the maximum for standard wheelchairs, when establishing architectural standards, which provide the physically handicapped with freedom of movement in buildings and other structures. However, a lesser width for a chair, or a means of temporarily reducing the full open width for passage through narrow doorways, would provide a definite asset for negotiating door openings in private homes or other structures, which do not conform the new standards.

(e) TURNING ABILITY: The turning radius should be held to a bare minimum. Turning should be accomplished without damage to floors or carpets. Chair must be able to negotiate an L or a U type stair landing. Ability to negotiate an L-type landing as small as 3 ft by 3 ft would be considered an asset.

(f) CLIMBING ABILITY: The chair should be able to negotiate street curbing and any stair with average height risers and depth tread that is found in office buildings and homes. Ascent should be performed in presence of litter and moisture without damaging stair treads and risers. Any transition or adjustments required between normal and ascending and descending functions must be accomplished in a minimum time to eliminate delays at street curbing and stairways.

(g) PROPULSION SYSTEM: The chair may be propelled by the occupant or by a motorized unit; however, total cost, weight, and collapsibility requirements are the same for either type of drive. If self-driven, minimum arm strength of 10 pounds may be used in devising drive mechanism.

(h) CHAIR SAFETY: The presence of an attendant, though undesirable, is permissible during ascent or descent. However, the chair occupant should always be in a reasonable state of balance, so that no more than 15 to 25 pounds of weight will be transferred to the attendant. The chair must be "fail safe" to prevent uncontrolled descent in the event something happens to the occupant, attendant, or mechanism.

(i) COST: In order for the maximum number of handicapped persons to be able to purchase a stair-climbing wheelchair, every effort should be made to keep its cost at a bare minimum. The retail cost should be no more than $200. Current models of standard tubular frame folding wheelchairs are priced at approximately $150.

(j) OTHER CONSIDERATIONS: Chair should be capable of performing all its normal functions without undue jostling or jouncing of the occupant. Since most paraplegics have little control of the lower limbs, a method of retaining them in position is often required. The footrests should be movable in order to enable the occupants to raise themselves to a standing position on the floor.

(k) The chair should not require the installation of special ramps, mechanical contrivances, or electrical outlets in buildings.

The preceding example has a clear statement with a clear set of requirements. Therefore, an objective tree, a function structure, and even the market analysis may not require much effort because all necessary details and specific requirements are spelled out in the statement. This is not always the case; 80 to 90% of the time the statement given to the design engineer is very vague. For instance, it may simply say, "Design a stair-climbing wheelchair." In this case an objective tree, functional analysis, and market analysis, are essential to clarify the statement. The ability to write a good set of engineering specifications is proof that the design team understands the problem.

There are many techniques for generating engineering specifications. Hubka separates the properties affected by the constraints into categories based on operational, ergonomic, aesthetic, distribution, delivery, planning, design, production, and economic factors. Pahl and Beitz based the specifications on the fulfillments of the technical functions, the attainment of economic feasibility, and the observance of safety requirements for the human and environment.

The approach that will be followed in this book is to develop a set of requirements based on the performance specification method and the quality function deployment method. In the performance specification method, the specification defines the required performance instead of the required product. It also describes the performance that a design solution has to achieve, but not any particular physical component that may constitute a means of achieving that performance. In the quality function deployment method, the attributes of a required performance are translated into engineering characteristics.

6.2 PROCEDURE FOR PERFORMANCE SPECIFICATION METHOD

The following are the steps in producing a set of specifications using the performance specifications method:

1. *Consider the different levels of generality of solution that might be applicable:* Specifications that are set at too high a level of generality may allow inappropriate solutions to be suggested. However, too tight specifications may remove all of the designer's freedom and creativity for the range of acceptable solutions. This level of limitation may also be connected with the definition of the customer you are dealing with, because the customer may also limit the class and the set of specifications. For example, designing a space shuttle puts the designer on a tighter limit of specification. Another example that demonstrates this point involves the design of an air jet for civilian use. Such a design has a different set of requirements than designing an air jet for military use. Recognizing the customer's needs is an important factor in defining the range for the set of requirements. The level of generality could be reserved through the development of the objective tree.

   Different types of generality levels can be listed from the most general to least general and may be demonstrated as

   a. Product alternatives
   b. Product types
   c. Product features

   Let us consider an example to illustrate these levels. Suppose that the product in question is a domestic aluminum can disposal. At the highest level of generality, the designer would be free to propose alternative ways of disposing of aluminum cans, such as crushing, melting, tearing, and chemical disposing. There might even be freedom to move from the concept of one can to multican, or from domestic to factory. At the intermediate level, the designer would have much more limited freedom and might only be concerned with different types of can crushers, such as foot operated or automatic can crushers. At the lowest level of generality, the designer would be constrained to consider only different features within a particular type of crusher, such as if it hangs on the wall or is self-standing, processes one or multicans in one step, and so on. Thus at this level of developing specifications, designers need to review the need statement and objective tree and analyze where the objective tree stands with respect to the mentioned levels.

2. *Determine the level of generality at which to operate:* In general, the customer/client determines the level of generality at which the designer may operate. For example, the customer may ask the designer to design a safe, reliable aluminum can crusher. Here, the customer has already established the type of product, and constraints are placed on the type of aluminum can disposable unit: It must be a crusher. Hence, the set of requirements will follow for a can crusher, not a disposable can unit.

3. *Identify the required performance attributes:* For clarifying the customer statement, objective trees, function analysis, and market analysis are used. Objective trees and functional analysis determine the performance attributes of the product. Some of the requirements in the customer statement are "must meet" (called demands) and others are "desirable" (called wishes). Solutions that do not meet the demands are not acceptable solutions, whereas wishes are requirements that should be taken into consideration whenever possible. Pahl and Beitz developed a checklist that can be used to derive attributes along with objective trees and function analysis (Figure 6.1). Once the attributes are listed, distinguish whether the attribute is a demand or a wish, and then tabulate the requirements.

4. *State succinct and precise performance requirements for each attribute:* The attributes that can be written in quantified terms must be identified. The performance limit should be attached to the attribute. The limits may be set by the customer statement or by federal and government agency standards. These attributes may include, but are not limited to, maximum weight, power output, size, and volume flow rate. For presenting the specifications, use a metric-value combination (i.e., size less than $1\ m^3$).

   For wishes and demands that do not have defined value attached to the metric, we use the quality function deployment method. Section 6.4 describes the quality function deployment method.

Example: Wheelchair Retrieval

Design a wheelchair retrieval unit to assist nurses in situations where a patient is taking a walk with his nurse and then in about 100 feet feels tired. The nurse must remain with the patient to support him and should be able to use one hand to activate the wheelchair retrieval unit. Design the wheelchair retrieval unit with the following requirements:

- Unit dimension within $1\ ft^3$
- Reach the patient in 1 minute
- Responsive within 100 ft away from the nurse

Main Headings	Examples
Geometry	Size, height, breadth, length, diameter, space, requirement, number, arrangement, connection, extension
Kinematics	Type of motion, direction of motion, velocity, acceleration
Forces	Direction of force, magnitude of force, frequency, weight, load, deformation, stiffness, elasticity, stability, resonance
Energy	Output, efficiency, loss, friction, ventilation, state, pressure, temperature, heating, cooling, supply, storage, capacity, conversion
Materials	Physical and chemical properties of the initial and final product, auxiliary materials, prescribed materials (food regulations etc.)
Signals	Inputs and outputs, form, display, control equipment
Safety	Direct safety principles, protective systems, operational, operator and environmental safety
Ergonomics	The man–machine relationship, type of operation, clearness of layout, lighting, aesthetics
Production	Factory limitations, maximum possible dimensions, preferred production methods, means of production, achievable quality and tolerances
Quality Control	Possibilities of testing and measuring, application of special regulations and standards
Assembly	Special regulations, installation, siting, foundation
Transport	Limitations due to lifting gear, clearance, means of transport (height and weight), nature and conditions of dispatch
Operation	Quietness, wear, special uses, marketing area, destination (for example, sulphurous atmosphere, tropical conditions)
Maintenance	Servicing intervals (if any), inspection, exchange and repair, painting, cleaning
Recycling	Reuse, reprocessing, waste disposal, storage
Costs	Maximum permissible manufacturing costs, cost of tooling, investment and depreciation
Schedule	End date of development, project planning and control, delivery date

Figure 6.1   Checklist for drawing up a requirement list.

Attention must be paid to appearance. The finished product should be marketed in two years' time. Manufacturing cost should not exceed $50 each at a production rate of 1000 per month.

The specifications are shown in Table 6.1.

This example is a specification for a wheelchair retrieval unit that can be operated with one hand and help a nurse who is supporting a patient to retrieve the wheelchair. The initial statement has been considerably expanded. The range of users has also been taken into account, as have safety considerations. The demands (D) and wishes (W) have been distinguished.

TABLE 6.1	Specification Table for a Wheelchair Unit	
D or W	**Specification**	**Importance (1–10)**
D	Dimension $1 \times 1 \times 1$ ft^3	10
D	Reach patient in 1 min	10
D	Water resistant	10
D	Stable	10
D	Ability to support patient weight	10
D	Ability to provide transportation & seating	10
D	Easy to operate	10
D	Low maintenance	10
D	Durable	10
D	Safe	10
W	Stops quick	7
W	Comfort	6
W	Easy to control	7
W	Low power drain when not in operation	5
W	Minimum amounts of parts	6
W	Lightweight	4
W	Compact/foldable	3
W	Low production cost	7
W	Operate by itself	3
W	Small turning radius	5
W	Maneuver through obstacles efficiently	8

6.3   SPECIFICATION TABLE: AUTOMATIC CAN CRUSHER

In the design process, after the designers have created the objective tree and the function tree, it is important that the designers complete a table of specifications. Designers must decide on the level of generality of the specifications. Market analysis results should be reviewed at this time. From the need statement for the aluminum can crusher, the specifications that *must* be met are as follows:

1. The design of the crushing mechanism is not to exceed $1.5 \times 1.5 \times 1$ ft^3 in the total size.
2. The can crusher must have a continuous feed mechanism.
3. The can must be crushed to $\frac{1}{5}$ of its original volume.
4. The device must operate safely. Children will use it.
5. The device is fully automatic.

The designers came up with specifications that were ranked as wishes. Not all wishes have the same level of importance. Thus, the design team gave a number for the rank of wishes. Figure 6.2 shows the specification table that the design team came up with for the automatic can crusher.

SPECIFICATION TABLE

Changes	D or W	Requirements	Importance (1 = Low, 10 = High)
	D	dimensions 1' x 1.5' x 1.5'	
	D	continuous can feed mechanism	
	D	crush to 1/5 original volume	
	D	can be operated safely by elementary school students	
	D	stand alone unit	
	D	total manufacturing cost <= $200,000	
	D	ability to mount to various surfaces	
	D	automatic removal of crushed crushed cans	
	D	enough force to crush cans ~30 lbs ~72 J	
	D	wiring kept away from moving parts	
	W	machine rendered inoperable when opened	10
	W	internal parts totally enclosed and safe from ambient environment	10
	W	crushing mechanism humanly inaccessible from feeder and dispenser	10
	W	runs on 110V standard US house outlet	9
	W	retails for < $50.00	3
	W	variable length/extractable cord	3
	W	enough force to crush glass, plastic	5
	W	internal parts safe from liquid damage	8
	W	large storage of crushed cans	3
	W	crushes plastic, glass and aluminum containers	2
	W	ability to crush various sizes of containers	3
	W	easily accessible kill switch	10
	W	large can capacity loader	4
	W	crush many cans/min.	8
	W	low noise output	7
	W	inconspicuous	5
	W	easy to disassemble	3
	W	easy access to clear jammed cans	6
	W	little heat produced	6
	W	no sharp corners	8
	W	easy to start	9
	W	starts up immediately	7
	W	long running capabiliy	9
	W	stops easily and immediately	8
	W	ability to stop in mid operation	10
	W	low vibration	8
	W	low peripheral force	8
	W	shock absorption	8
	W	high efficiency engine	9
	W	consolidate mechanical functions	5
	W	utilize gravity in design	8
	W	drain for residual liquid from machine	2
	W	easy cleaning	3
	W	light weight	3
	W	portable	4
	W	high material strenght	9
	W	durable refuse container	4
	W	built from a polymer	2

Figure 6.2   Specification table for automatic can crusher.

W	housing constructed from a formed polymer	2
W	extra wiring insulation	6
W	low loading height	7
W	easily accessible interior	7
W	small force required to depress the switches	9
W	safety stickers	9
W	low standby power drain	5
W	green light to indicate O.K to load	3
W	red light to indicate the crushing mechanism is in operation	3
W	yellow light to indicate improper use of the machine	3
W	reset button to reset machine after kill switch has been used	9
W	can counter	2
W	flip open lid	1
W	container to hold refuse liquid	2
W	no exhaust	8
W	no flying debris	10
W	automatically switches to standby power when not in use	3
W	operation steps sticker	9
W	colors available	1
W	stylish, fashionable, blends with surrounding	5
W	printable surface	1
W	receiving container on caster	3
W	no solvents able to hurt finish	6
W	parts easy to acquire	9
W	plexiglas window to view operation	1
W	no service required	7
W	weather proof	3
W	sealed bearings	7
W	limit number of tight tolerances	7
W	utilizes ground to stabilize	5
W	less than five assembly steps	5
W	ability to mount to various surfaces	5

Figure 6.2   Specification table for automatic can crusher (continued).

Note that the wishes are tabulated according to their importance value (1 to 10). The higher the value, the more important is the wish on the list. This technique is explored further in the following section.

6.4   QUALITY FUNCTION DEPLOYMENT METHOD

The quality function deployment (QFD) method was first developed in Japan in the mid-1970s and used in the United States in the late 1980s. Through the use of the QFD method, Toyota was able to reduce the cost of bringing a new car model to the market by more than 60%, and the time was reduced by about 33%. Steps in the QFD method are as follows:

1. *Identify the customer requirements in terms of product attributes:* You have done this through question/answer sessions with the customer and through discussion sessions with the design team to develop the objective tree. Market research, function analysis, and the performance specification method also help

in this effort. In cases where a company is launching/designing a new product where there is no particular customer at hand, gathering attributes could be done through direct interaction with prospective customers by conducting

a. Product clinics: Customers are quizzed in depth about what they like and dislike about a particular product.
b. Hall test (tests conducted in the same hall): Various competing products are arranged on display and customers are asked to inspect the products and give their actions.

In the QFD, consumers are the most important factor, and the design should be arranged in a manner to meet their satisfaction. The Kano model can be used to measure customer satisfaction. In the Kano model, customer satisfaction is measured by the product function, as shown in Figure 6.3.

In the Kano model, there are three different types of product quality that give customer satisfaction: basic quality, performance quality, and excitement quality. (The Kano model was developed by Dr. Noriaki Kano in the early 1980s.) With basic quality, customers' requirements are not verbalized because they specify assumed functions of the device. The only time a customer will mention them is if they are missing. If they are not fully implemented in the final product, the customer will be disgusted with it. If they are included, the customer will be neutral. An example is the requirement that a bicycle should have brakes. The performance quality refers to customers' requirements that are verbalized in the form that indicates the better the performance, the better the product. The excitement quality involves those requirements that are often unspoken because the customer does not expect them to be met in the product. If they are absent, customers are neutral. However, if the customers' reaction to the final product contains surprise and delight at the additional functions, then the product's chance of success in the market is high.

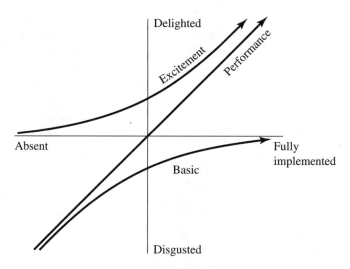

Figure 6.3   Kano model for customer satisfaction.

2. *Determine the relative importance of the attributes:* The demands are absolute requirements. The wishes, however, are not in the same order of importance. In the event where you as a designer are directly interacting with the customer, the ordering of the wishes could be done by asking the customer to rank these requirements in order and hence generate an ordered list of required wishes. As a group team working on a design project, you could rank these wishes in order or conduct a survey in the event that the product you are working on will be beneficial to a group of customers that are easily identified and accessed. Companies desiring to conduct such ordering of wishes use surveys. To distinguish numerically the importance of each wish, a weighing factor is used. There are different methods to allocate weighing factors for these wishes:

   a. Absolute measure for each wish individually, where each wish is rated from 1 to 10. Wishes that have the same importance are assigned the same value. The total measure is irrelevant.

   b. Relative measure, where a total of 100 is maintained and each wish is given a value according to its importance. The weighing factor in this event could be a percentage figure adding to 100% when adding all values.

   In many instances the customer will be more than one entity (For example, the brake pads for a car would have the drivers, the mechanic, and salespeople as essential customers who may view and evaluate attributes differently.) In such events the attributes will be evaluated in three different columns according to their importance. Thus all three customer categories will have to input the value for the weighing factor as they see fit. Students need to be aware of this strategy; however, in many design cases where students compose the design team and the instructor or an industry representative is the customer, it is reasonable to ask the customer and the design team to grade the wishes.

3. *Evaluate the attributes of competing products:* The objective here is to determine how the customer perceives the competition's ability to meet each of the requirements. Usually customers make judgments about the product in terms of comparison with other products. This step is very important because it shows opportunities for product improvement. The market analysis you have conducted should play a vital role in this step. In many instances students who do not conduct extensive market analysis find it hard to complete this step accurately.

4. *Draw a matrix of product attributes against engineering characteristics:* Attributes are usually vague and general; the characteristics are measurable, real, and quantitative. This step allows you to define your vague and ill-measured objectives. One technique is to list a few ways you could measure an objective. Thus you would brainstorm for metric-value combinations that are suitable in measuring an objective. For example, one of the objectives could be that the product is lightweight. Engineering characteristics to measure this objective can be listed as follows:

   a. Volume of the product should be less than [*value suitable to the product*].
   b. Mass of the product should be less than [*value suitable to the product*].
   c. Material density should be less than [*value suitable to the product*].

   Use an Excel sheet to place product attributes (from the specification performance table, objective tree, etc.). Make sure you will be able to distinguish between the demands and wishes. Use the first column for this listing. Use the first row to list the engineering characteristics.

5. *Identify the relationship between the engineering characteristics and the product attributes:* Each cell in the table created in the Excel spreadsheet represents how an engineering characteristic relates to the customer requirement. Many characteristics will measure more than one customers' requirements. The strength of the relationship can vary, with some engineering parameters providing strong measures for customers' requirements and others providing no measure at all. The relation is conveyed through numerical values such as the following:
   a. 9 = strong relationship
   b. 3 = medium relationship
   c. 1 = weak relationship
   d. blank = no relationship
   e. x = demand

6. *Identify any relevant interactions between engineering characteristics:* Engineering characteristics may have relationships between each other. For example, a powerful engine is also likely to be a heavier engine. This interaction is added as a roof to the matrix. This is called the house of quality. A similar procedure is to list the engineering characteristics in both the first row and first column of an Excel sheet.

7. *Set target figures to be achieved for the engineering characteristics to improve over competitors:* The market analysis may be important at this stage to identify the market limits.

Example: Controlled Vents

The design team in charge of designing an adjustment to current vents such that the vents could be opened and closed remotely from a centralized location in the house has developed a house of quality as shown in Figure 6.4. In this figure the specifications are listed in the left column. The designers were the only people to evaluate the specifications. The designers developed engineering characteristics to measure the specifications as shown in Figure 6.4.

Example: Splashguard[1]

This example is of a relatively simple product. Yet it illustrates that considerable effort may be necessary in designing to satisfy customer requirements even for a simple product. The design team identified three important customers for the product: the rider, the mechanic, and the marketing. The mechanics are considered customers because they will sell, and perhaps install, the splashguard. Marketing is added to the customer list to ensure capture of production and marketing needs. Figure 6.5 shows a mountain bicycle and the customer's list of requirements for the splashguard. Figure 6.6 shows the house of quality that is developed based on the QFD method for a splashguard.

---

[1]Adapted from Ulman.

	designer	noise <30 db	profit 25%	parts <15	steps to operate <5	maintenance cost <$20 anually	disassembly <15 minutes	force <15 newtons	production cost <$200 per unit	time to open vent <5 seconds	cost to use <$5 month
easy to operate	9				9						
low maintenance	3		1	9		9	3			1	
few parts	5			9	3		3				
safe for user	11	3			3						
safe for environment	9	3									3
enough power to do job	4				3			9		3	
fits vents universally	4		9								
low noise	2	9		3	3						
low operation cost	4		3			1			3		9
inconspicuous	1	3		3							
profitable	8		9	3		3			9		3
easy to disassemble	2			3			9				
low vibration	2	3			3			9		3	
long lasting	2			9	3			3		3	
small actuating force	1							3		3	
time in use small	2	3		3	9					9	
inexpensive material	4		9			3			9		
inexpensive to consumer	4		1	3		3			9		3
conserve energy	2			1	3			3			9
easy to repair	2			9		3	9				
quality materials	3	3									
different settings	4		3		3		3				3
remotely operated	12		9	9			1		3		
Total	100										

Figure 6.4   House of quality for controllable vents.

Preliminary list of customers'
requirements for the splashguard

Riders' and bike shop mechanics' requirements
    Keeps water off rider
    Is easy to attach
    Is easy to detach
    Is quick to attach
    Is quick to detach
    Won't mar bicycle
    Won't catch water/mud/debris
    Won't rattle
    Won't wobble
    Won't bend
    Has a long life
    Won't wear out
    Is lightweight
    Won't rub on wheel
    Is attractive
    Fits universally
    If permanent piece on bike, then is small
    If permanent piece on bike, then is easy to attach
    If permanent piece on bike, then is fast to attach
    If permanent piece on bike, then is noninterfering
    Won't interfere with lights, rack, panniers, or brakes

Company management requirements
    Captial expenditure is less than $15,000
    Can be developed in 3 months
    Can be marketable in 12 months
    Manufaturing cost is less that $3
    Estimated volume is $200,000 per year for 5 years

Figure 6.5   Requirements for splashguard.

	Mechanic	Marketing	Rider	Water hitting rider (%)	Steps to attach (#)	Time to attach (sec)	Steps to detach (#)	Time to detach (sec)	Number of parts (#)	Weight (g)	Customers finding it visually appealing (%)	Color available (#)	Bikes it fits (%)	Upward release force (N)	Sales price ($)	Whale Tail	Norco	Raincoat
Functional performance																		
Keeps water off rider	1	1	7	9										3		1	4	2
Fast to attach	5	4	8		3	9			3	1						1	4	3
Fast to detach	9	5	10				3	9	3	1						2	4	3
Can attach when dirty	7	13	12		3	3										3	3	2
Can detach when dirty	11	12	13				3	9								3	3	2
Human factors																		
Easy to attach	4	6	9		9				3	1						1	3	3
Easy to detach	10	7	11				9		3	1						1	4	3
Looks fast	2	10	2								9					4	2	2
Color matches bike	12	11	5								3	9				3	2	2
Interface with bike																		
Fits bike	3	3	3										9			3	2	4
Does not mar bike	8	8	6											1		3	1	4
Lightweight	6	9	4							9						3	3	4
Competitive sales price	13	2	1												9	2	3	1
Whale Tail				25	5	25	2	5	6	130	75	5	94	5	12			
Norco				0	3	5	1	3	2	140	65	1	65	15	15			
Raincoat				30	3	10	3	10	1	100	35	4	100	0	20			
Target				0	1	2	2	3	2	130	85	5	95	5	10			

Figure 6.6 House of quality for the splashguard.

6.5 HOUSE OF QUALITY: AUTOMATIC CAN CRUSHER

The design team has taken the wishes and objectives from the specification table and the objective tree and placed them into the house of quality. Because of the large number of wishes they listed on their table, they were unable to show the house of quality in one figure. However, four tables are shown that represent the house of quality (Figures 6.7 and 6.8). On the first table, the wishes are placed with the

House of quality for automatic can crusher

Customer requirement	probability of injury < 0.1 %	weight < 30 lb	sales price < $50.00	number of parts < 100	dimensions (inch)	crushing force > 30 lb	people able to use > 5 yrs	Manufacturing cost <$200	Steps to operate (1)	maintenance cost < $10 annually	efficiency rating > 95 percentile	internal parts enclosed (100%)
easy access to clear jammed cans	3			6		3	3			3	6	
crushing mechanism humanly inaccessible	9	1	3	3	3		9	3				9
machine rendered inoperable when opened	9			1	6		9	9				1
easily accessible kill switch	9		3	7			9	3	3			
ability to stop in mid operation	9			1	3		9	6	1			
no flying debris	9						9			1		6
runs on 110V standard U.S house outllet			6	3		3	9	9	6		9	
easy to start			3				9	6	6			
long running capability			6	1			7			9	9	
high efficiency engine			6	1		9		9		9	9	
high material strength			9	6		3		9			6	
small force required to depress the switches	9						9	3				
safety stickers	9						9	2	2			
machine reset button after kill switch activated	9			6			9	3	2			
operation steps sticker				6			9	3	9			
parts easy to acquire			6				3	9		9		
internal parts safe from liquid damage	2		4	4				3		9	6	9
crush many cans/min.			9	6	1	9	3	9			6	
no sharp corners	9						9	6				
stops easily and immediately	9			3			9	6	1	6		
low vibration		6		3	3	6		3		1		
low peripheral force		6		3	3	6		3		1		
shock absorption		6		3	3	6		3		1		
utilize gravity in design		9	3	6	9			9			1	
no exhaust	9		1	1			9					
low maintenance cost			6	6			8	3		9	6	1
low noise output	9		4	3			6	6				9
starts up immediately			3	5				3			6	
low loading height	6		1		9		9					
easily accessible interior	3		3				6	3		3		3
no service required			9	2			6			9	9	
sealed bearings								3				6
limit number of tight tolerances												
low operation cost			9	3			6	3		6	6	
easy access to clear jammed cans	6		3				3	3		6	3	3
little heat produced	1			1		2					9	
extra wiring insulation	6	1		1				1				
enough force to crush glass, plastic						9	6	9				
inconspicuous			6			6	6	3				

Figure 6.7   House of quality for automatic can crusher.

consolidate mechanical functions	3		9				6	2		9			
low standby power drain		1				3				9			
aesthetically pleasing/blends with surrounding		9	2	6		8	6						
utilizes ground to stabilize			1										
less than five assembly steps		3	6			3	1						
ability to mount to various surfaces	3	6		3		3	3						
large can capacity loader	3	6		3		3	3	6					
portable	9	6		3		6	3						
durable refuse container						3							
retail for < $50.00		9				9	9						3
variable length/retractable cord		3				2	1						
large storage of crushed cans	1	4		3		2	1	6					
ability to crush various sizes of containers		9	3	8		6	9						
easy to disassemble		3	6			3	3				6		
easy cleaning		6				3	1				3		6
green light to indicate OK to load	9	6				9	2						
red light to indicate the crushing mechanism is in operation	9	6				9	2						
yellow light to indicate improper use of the machine	9	6				9	2						
automatically switches to stadby power when not in use										6			
receiving container on caster	2												
weather proof		1	3			3					6		9
crushes glass, plastic and aluminum containers		9	6	3	9	7	9						
drain for residual liquid from machine	1		1			1					6		
built from a polymer	6	6				6							
housing constructed from a formed polymer	6	6				6							
can counter		3	2			1	2						
container to hold refuse liquid	2												
flip open lid													
colors available		9				3							
paintable surface		8				2							
Plexiglas window to view operation		8				2							6

Figure 6.7   House of quality for automatic can crusher (continued).

engineering characteristics, while on the second table, the engineering characteristics are evaluated against each other. The engineering characteristics are obtained by including a metric, a value, and a unit. Numerical values between 1 and 9 are assigned between each wish and engineering characteristic to rank the correlation between them; 1 is a very low correlation and 9 is a very high correlation. Blank spaces correspond to no correlation.

House of quality for automatic can crusher. Engineering characteristics and relationship matrix:

storage of crushed cans (60 cans)	loader capacity > 30 cans	crush >= 15 cans/min.	crush >= 1.2 × 10⁻² m³/min	crush >= 0.57 kg/min.	noise output < 30 dB	Starts < 10 sec	runs > 2hrs at a time	stops < 5 sec	vibration magnitude < 5 mm	vibrations < 4/sec	withstand <= 250 N	maintenance < 4hrs/yr	# colors (6)	lifespan > 4000 hrs
		6	6	6								3		3
								1						
								9						
								9						
		1	1	1				9						
		6	6	6	3		1							
						9								
	1	2	2	2	3		9		2	1		9		9
		6	6	6	2				2	2		9		9
		1	1	1								6		9
						6								
												9		
												6		9
6	9	9	9	9	9	1	1		3	3				3
								9						
		9	9	9	9				9	9				3
		9	9	9	9				9	9				3
		9	9	9	9				9	9				3
					3				1	1				3
					1									
							6		6	6	7	9		9
		1	1	1	9		1		3	3				
		1	1	1		9								
	9													
		3	3	3				1				9		6
						3					6	9		9
												9		6
		3	3	3										
						6						9		3
		3	3	3			6					6		4
						3			1	1		1		4
														2
		9	9	9								9		
6	6								6	6			9	

Figure 6.7 House of quality for automatic can crusher (continued).

109

Figure 6.8 Engineering characteristics.

Column / row key for the correlation matrix:

1. probability of injury < 0.1 %
2. weight < 30 lb
3. sales price < $50.00
4. number of parts < 100
5. noise output < 40 dB
6. dimensions (inch)
7. cans crushed (20/min)
8. crushing force > 30 lb
9. people able to use > 5 yrs
10. manufacturing cost <$200
11. steps to operate (1)
12. maintenance cost < $10 annually
13. efficiency rating > 95 percentile
14. internal parts enclosed (100%)
15. storage of crushed cans (60 cans)
16. loader capacity > 30 cans
17. starts < 10 sec
18. runs > 2hrs at a tims
19. stops < 5 sec
20. vibration magnitude < 5 mm
21. vibrations < 4/sec
22. withstand <= 250 N
23. maintenance < 4 hrs/yr
24. # colors (6)
25. lifespan > 4000 hrs

Characteristic	1	2	3	4	5	6	7	8	9	10	11	12	13	14	15	16	17	18	19	20	21	22	23	24	25
lifespan > 4000 hrs		9	9									3	9		6	6	6	6					9		9
# colors (6)		9												6										9	9
maintenance < 4 hrs/yr		9	9					3				9	9	9									9		9
withstand <= 250 N																						9			
vibrations < 4/sec	3	3	3					6				3	6								9				6
vibration magnitude < 5 mm	3	3	3					6				3	6							9					6
stops < 5 sec.													6						9						6
runs > 2 hrs at a tims							6					6	6		6	9		9							6
starts < 10 sec		6	3						9				6				9								
loader capacity > 30 cans	6						6	9				6			6	9	9								
storage of crushed cans (60 cans)	3						6					3	6		9	6		6							
internal parts enclosed (100%)									6					9										9	
efficiency rating > 95 percentile		9	9				9	9	6			6	9	9		6	6	6	6	6	6	3	6		9
maintenance cost < $10 annually		6	9						6			9	9				6	6		3	3		9		
steps to operate (1)	3		6						9		9	6													3
manufacturing cost <$200	3	3	9	9		3			9						6	3	6						6		
people able to use > 5 yrs	9		9	9				9				9	6	6			9								
crushing force > 30 lb	3		3				9	9					9							6	6				
cans crushed (20/min)		9		6			9	9				9			9	9	9	6					3		
dimensions (inch)	1			9			3						6	6											
noise output < 40 dB	9	6	3	9		6		9					6								9				
number of parts < 100	9	6	3	9	3			9		9	9				3								9	9	
sales price < $50.00	6	9	3	6		9	3	9	9	6	6	9			6					3	3		9	9	9
weight < 30 lb	9	6		1		3	3	3						3	6					3	3		3	3	
probability of injury > 0.1 %	9	6	9		3	9	3	3	3					9	6					3	3		3	3	

6.6   PROBLEMS

6.6.1   Team Activities

1. Given the need statement "Design a domestic heating appliance," generate different levels of generality of solutions.
   a. Alternatives
   b. Types
   c. Features
2. Generate engineering characteristics for the following:
   a. Easy to hold
   b. Cost effective
   c. Safe for the environment
   d. Easy to maintain
3. Consider the statement you worked on previously during an in-class team activity. By looking at your objective tree, generate a list of specifications and house of quality for the statement.

   If you drive in the state of Florida, you notice some traffic congestion due to trees being trimmed on state roads and freeways. Assume that the Florida Department of Transportation is the client. Usually, branches 6 inches or less in diameter are trimmed. The allowable horizontal distance from the edge of the road to the branches is 6 ft. The material removed from the trees must be collected and removed from the roadside. To reduce the cost of trimming, a maximum of only two workers can be assigned for each machine. The overall cost, which includes equipment, labor, etc., needs to be reduced by at least 25% from present cost. The state claims that the demand for your machines will follow the price reduction (i.e., If you are able to reduce the cost by 40%, the demand will increase by 40%.) Allowable working hours depend on daylight and weather conditions.

6.6.2   Individual Activities

A specification consists of a metric and a value. For example, "average time to assemble" is a metric, while "less than 75 seconds" is the value of this metric. Note that the value may take on several forms, including a particular number, a range, or an inequality. Values are always labeled with the appropriate units. Together, the metric and value form a specification. The product specifications are simply the set of the individual specifications.

1. Develop a list of specifications for the following objectives. The table should include metrics, the corresponding values (assume a value), and units.
   a. Easy to install
   b. Reduces vibration to the hand
   c. Is not contaminated by water
   d. Lasts a long time
   e. Is safe in a crash
   f. Allows easy replacement of worn parts

2. List a set of metrics corresponding to the need that a pen writes smoothly.

3. Devise a metric and a corresponding set for the need that a roofing material lasts many years.

4. How might you establish precise and measurable specifications for intangible needs such as, "The front suspension looks great?"

5. Can poor performance relative to one specification always be compensated for with high performance on the other specifications? Give examples.

6. Develop a house of quality for the following need statement: "We have a mountain-sized pile of wood chips that we want processed into home use fire logs. We are located in Tallahassee, Florida. We can have a continuous supply of wood chips throughout the year. We should be able to produce 50 fire logs per minute. The shareholders require us to have a large profit margin, and our prices should be lower than our competition for the same size logs. Needless to say, the logs should produce enough heat to keep our customers faithful and have environmentally clean exhaust."

7. Use a Web search engine to identify a list of companies that utilize QFD. Describe what they apply QFD for and list differences and similarities in the steps developing QFD.

6.7   SELECTED BIBLIOGRAPHY

1. Akao, Y. *Quality Function Deployment: Integrating Customer Requirements into Product Design* [Translated by Glenn Mazur]. Productivity Press, 1990.

2. Clausing, D., and Pugh, S. "Enhanced Quality Function Deployment," Design and Productivity International Conference, Honolulu, HI, Feb. 6–8, 1991.

3. Cohen, L. *Quality Function Deployment: How to Make QFD Work for You.* Reading, MA: Addison-Wesely, 1995.

4. Daetz, D. *Customer Integration: The Quality Function Deployment (QFD) Leader's Guide for Decision Making.* New York: Wiley, 1995.

5. Day, R. G. *Quality Function Deployment: Linking a Company with Its Customers.* ASQC Quality Press, 1993.

6. Dean, E. B. "Quality Function Deployment for Large Systems," Proceedings of the 1992 International Engineering Management Conference, Eatontown, NJ, October 25–28, 1992.

7. Hauser, J. R., and Clausing, D. "The House of Quality," *The Harvard Business Review*, May–June, No. 3, pp. 63–73, 1988.

8. Hubka, V. *Principles of Engineering Design.* London: Butterworth Scientific, 1982.

9. King, B. *Better Designs in Half the Time: Implementing Quality Function Deployment in America.* GOAL/QPC, Methuen, MA, 1989.

10. Mizuno, S., and Akao, Y. *QFD: The Customer-Driven Approach to Quality Planning and Deployment* [Translated by Glenn Mazur]. Asian Productivity Organization, 1994.

11. Moen, R., Nolan, T., and Provost, L. P. *Improving Quality Through Planned Experimentation.* New York: McGraw-Hill, 1991.

12. Nakui, S. "Comprehensive QFD." *Transactions of the Third Symposium on Quality Function Deployment*. Novi, MI, June 24–25, pp. 137–152, 1991.

13. Pahl, G., and Beitz, W. *Engineering Design: A Systematic Approach*. New York: Springer-Verlag, 1996.

14. Pugh, S. *Total Design*. Reading, MA: Addison-Wesley, 1990.

15. Shillito, M. L. *Advanced QFD: Linking Technology to Market and Company Needs*. New York: Wiley-Interscience, 1994.

16. Ulman, D. G. *The Mechanical Design Process*. New York: McGraw-Hill, 1992.

17. Vidosic, J. P. *Elements of Engineering Design*. New York: The Ronald Press Co., 1969.

7 Developing Concepts

A design team is required to design a retrieval unit for wheelchairs to assist nurses performing walking activities with patients. The design team for this wheelchair retrieval unit develops a function analysis for the unit and then creates a morphological chart. (Copyright Siedes Preis/Getty Images/PhotoDisc.)

7.1 OBJECTIVES

By the end of this chapter, you should be able to

1. Generate a morphological chart.
2. Understand a better way to arrange design steps.

So far we have developed methods to understand and define customers' needs as clearly as possible. We used objective trees to expand on the needs and determine the various levels of goals. We used functional trees to address the different functions that need to be carried out in a hierarchical manner, to satisfy the needs. Then we drew up the specification chart based on what we needed to achieve and what we needed to do to achieve the objectives. The specifications that must be satisfied were classified as "must", and those that will simply enhance the product quality as "desired." However, specifications describe what needs to be achieved in a measurable quantity rather than describing a qualitative desire expressed at the objective stage. The next step is to come up with more than one artifact that will satisfy the needs of the customer, taking into consideration the desires and wishes of the specifications. Now you have reached a stage where concepts need to be generated based on what you have accomplished in the previous steps in the design process. These concepts represent multiple solutions to the problem.

As was discussed in previous chapters, the goal of the design process is to develop artifacts that satisfy a need. Although the majority of designs can be considered as redesigns or modifications of existing products, these modifications need to be new and unique. The systematic process of design that you have followed stresses not having your mind set on a specific solution. Thus we have considered the process as producing several alternatives; later you will find that the process converges to one unique solution. Several design methodologies stress the concept of brainstorming as a model to generate several alternatives. In the presented design process, brainstorming plays an important role in generating alternatives for each function of the proposed solution. In this stage brainstorming is important, however you can't use brainstorming without going through the process (i.e., at each stage you will need brainstorming but you can't brainstorm without going through the process.)

7.2 BRAINSTORMING

The basic concept of brainstorming is to generate a large quantity of ideas. Research shows that the more ideas, the higher the quality of the desired product. A brainstorming session requires participants to be prepared to offer all ideas, including seemingly silly ideas. In many cases silly ideas lead to genuine and creative ideas. Designers are encouraged to include nonengineers in the brainstorming sessions; often the best ideas come from people who do not have expert opinions on the area or subject at hand. Marketing and psychology experts are often included in brainstorming design sessions.

In the systematic design process, you have used brainstorming in generating objective tree and function analysis. Concept generation is an important step; thus we focus on mechanisms and tools that can facilitate the improvement of the design output. Remember the iterative nature of the design process. If you have gathered information that you must update, you can do so at this stage.

7.2.1 Mechanism of Brainstorming Session

Participants must be prepared to be highly involved and active in the session. The topics of the brainstorming session need to be specific and stated clearly. Because of the energy level needed for such sessions, it is recommended to provide up to an hour for each topic/session. A set of rules has been developed for brainstorming sessions:

1. No criticism of ideas during the session. One way to prevent the criticism is to disallow discussion during the session. Present the ideas in brief and short format. One of the participants in the group will need to collect all ideas or use a sticker for each idea.
2. Wild, silly, and crazy ideas are welcomed. Such ideas may help others generate solid ideas or help maintain a fun and humorous environment.
3. Generate as many ideas as possible. One way to do so is to appoint a facilitator who will keep rotating the turns without allowing break time. Keep the competition spirit alive. Some people produce more when competing with others in a group.
4. Adding to or improving presented ideas is welcomed. This will help maintain a stream of ideas.

7.3 CREATIVITY

In many design books there is a strong connection between design and creativity. Some research concludes that design is a creative process by nature. In this section a definition of creativity in relation to design is introduced.

Ned Herrmann, author of *The Creative Brain*, defines creativity as follows: "Creativity in its fullest sense involves both generating an idea and manifesting it—making something happen as a result. To strengthen creative ability, you need to apply the idea in some form that enables both the experience itself and your own reaction and others' to reinforce your performance. As you and others applaud your creative endeavors, you are likely to become more creative."

Lumsdaine defines creativity as "Playing with imagination and possibilities, leading to new and meaningful connections and outcomes while interacting with ideas, people and environment."

Kemper has the following to say about creativity: "The word creativity is a rather curious state: Nearly everyone can recognize creativity instantly, yet no one can define the word in a fully acceptable manner. A companion word, invention, causes even more trouble. Every layperson is instantly confident of the word invention and knows exactly what it means; such a person can think of countless inventions: The electric light, the safety pin, even the atomic bomb. It is only Patent Office examiners and United States Supreme Court justices who believe there is a problem in defining what an invention is."

In these definitions we see that creativity is associated with generating ideas. We need to ask if creativity is an individual quality or can be associated with a group or team. Further, the more important question is if creativity is natural or can be taught. If creativity is natural, then creativity can be categorized in each individual as either present or absent. However, research has found that there is a degree of creativity in each individual that depends on the cognitive style, personality, and the creative outcome produced. Psychologists suggest that the level of creativity can be determined either statistically or assessed by experts.

Research has found that groups perform better on creative problem-solving tasks. It has been said that two heads are better than one, which may be also applied to creativity. Brainstorming is the best-known and most widely used technique for idea generation in groups. However, for brainstorming sessions to outperform individuals in generating ideas, participants must (1) have some social relationships (e.g., as students you identified with each other the first day you walked into the classroom); (2) have used some of the idea generated (you have done this through the objective tree and function tree); (3) have some technical experience pertinent to the problem (although not as important but preferable); and (4) have worked some tasks interdependently.

7.3.1   How to Increase Your Level of Creativity

The author John Steinbeck once said, "Ideas are like rabbits. You get a couple and learn how to handle them, and pretty soon you have a dozen." The following will help you become more creative:

1. *Know your thinking style:* Since early history humans have been trying to understand how the brain works. Herrmann developed a metaphorical model of the brain that consists of four quadrants. Although all of us are using all four quadrants, some individuals may have more use of certain quadrants of the brain. The following is a listing of these quadrants and their characteristics:

   a. Upper left: The characteristics of this quadrant are analytical, logical, quantitative, and fact based.
   b. Lower left: The characteristics of this quadrant are organized, planned, detailed, and sequential.
   c. Upper right: The characteristics of this quadrant are holistic, intuitive, synthesizing, and integrating.
   d. Lower right: The characteristics of this quadrant are emotional, social, and communicative.

   In team work, you may encounter different thinking styles. A good team is one that represents a full brain. You can train yourself to have all quadrants function at the same power or increase activities toward utilizing more of a specific quadrant by:

   a. Identifying the weakness; and
   b. Attacking problems that require the utilization of a weak quadrant.

2. *Use visual imagery:* Einstein asserted that imagination is more important than knowledge for knowledge is finite whereas imagination is infinite. Bernard Shaw said, "You see things and say why? But I dream things that never were and I say why not?" The inner imagery of the mind's eye has played a central role in the thought processes of many creative individuals. Albert Einstein ascribed his thinking ability to imagining abstract objects. Kekule discovered the all-important structure of the benzene ring in a dream. Most visual thinkers clarify and develop their thinking with sketches. Drawing not only helps them to bring vague inner images into focus; it also provides a record of the advancing thought stream. Sketches expand the ability to hold and compare more information than is possible from simple recall from the brain. Sketches are usually extensions of thought, and they provide a conference table to talk to oneself. Distinguishing between the state of sleep and awake and between visions and perceptions has been studied since ancient history. According to researchers in visual imagery, there are three stats of visual imagery:

   a. Dharana, in which you simply focus on a given place or object.
   b. Dhydana, in which you strengthen the focus of attention.
   c. Samadhi, in which you experience a union fusion with the object and cannot distinguish between self and object.

   As students, you have been practicing memorization since kindergarten; visualization will help enhance your memory as each image stores much more information than words. The recollection of these images will enhance your idea generation.

3. *Reframing:* Einstein asserted that problems cannot be solved by thinking within the framework in which the problems were created. Reframing involves taking problems out of their frame and seeing them in a different context. It allows consideration of potentially valuable ideas outside current frames. The most common habits that limit the ability to change mental frames are as follows:

   a. Pursuit of perfection: Many people are under the false impression that working long and hard is sufficient to develop the perfect answer to the problem.
   b. Fear of failure: Resistance to change because others may perceive you as incompetent.
   c. Delusion of already knowing the answer: Once you find an answer, you do not look for others. Why look for another answer if this one works?
   d. Terminal seriousness: Many people are under the illusion that humor and serious idea generation do not mix.

4. *Humor:* It was said that "Men of humor are always in some degree men of genius." The dictionary definition of humor is "The mental faculty of discovering, expressing or appreciating ludicrous or absurdly incongruous elements in ideas, situation, happenings or acts." If you were to take the words *ludicrous* or *absurdly* out of the definition of humor, you would have a definition of creativity.

5. *Information gathering:* In the market analysis (Chapter 4), you were introduced to the power of gathering information on the problem statement. Gathering information will also enhance creativity; it will allow you to view ideas generated by other creative minds (written brainstorming). Patents can serve as an excellent source of ideas. However, it is difficult to identify the specific patent that contains the idea that you are looking for. There are two main types of patents: utility patents and design patents. Utility patents deal with how the idea works for a specific function. Design patents only cover the look or form of the idea. Hence, utility patents are very useful since they cover how the device works, not how it looks. Since there are about 5 million utility patents, it is important to use some strategies to hone in on the one that you may be able to use. Use a Web search, such as that provided by the Patent Office. Key word searches as well as patent numbers, inventors, classes, or subclasses are available.

The next obvious place to find information is in reference books and trade journals in the relevant area.

Another obvious information source is to consult experts. If designing in a new domain, you have two choices for gaining the knowledge sufficient for generating concepts. You either find someone with expertise, or you spend time gaining experience on your own. Experts are those who work long and hard in a domain, performing many calculations and experiments themselves to find out what works and what does not. If you cannot find or afford an expert, then the next best source is the manufacturer's catalog or manufacturer's representative.

In the systematic design process, creativity is utilized in all steps. The need for group creative problem solving is utilized in constructing the objective tree, deducing information from market analysis, developing function analysis, structuring a house of quality, and developing concepts. The mechanism of developing concepts is discussed in the following few sections.

7.4 DEVELOPING WORKING STRUCTURES

Once the various functions at various levels are known, it is important to find the principles at work for each of the functions and subfunctions. A working principle must be based on the physical effect needed to achieve the given function based on the flow of materials, energy, and information within each function. Each function may be achieved in a number of ways. One important tool that can be effectively used is a design catalog.

Figure 7.1 shows the schematic of a design catalog; each subfunction is shown in the left column and the many possible solutions in the rows. Figure 7.2 shows an extract from a classification scheme for a mechanical vegetable collection system. In some cases, there are as many as eight possible ways to achieve the same function. The general idea is to identify as many means possible to achieve the same functional requirement.

Possible Solution/Options Sub Function		1	2	...	j	...	m
1	F_1	O_{11}	O_{12}		O_{1jj}		O_{1m}
2	F_2	O_{21}	O_{22}		O_{2jj}		O_{2m}
:		:	:		:		:
i	F_i	O_{i1}	O_{i2}		O_{ijj}		O_{im}
:							
n	F_a	O_{n1}	O_{n2}		O_{njj}		O_{nm}

F = Function of sub function
O = Options
ij = i,j solution/or option

Figure 7.1   Basic structure of a classification scheme with the subfunctions of an overall function and associated solutions.

Another frequently occurring design situation involves using the same working principle but changing the geometry or material combination (FORM) to enhance the functional requirement.

7.5   STEPS TO DEVELOP CONCEPTS FROM FUNCTIONS

Step 1: Develop concepts for each function.
The goal of this first step is to generate as many concepts as possible for each of the functions identified in the decomposition. There are two activities here that are similar to each other. First, for each function, develop as many alternative options functions as possible. Second, for each subfunction, the goal is to develop as many means of accomplishing the function as possible. In the previous example, vegetable placing could be achieved using an conveyor belt, a rake, a rotating mover or simply by accumulating the vegetables. If there is a function for which there is only one conceptual idea, this function should be reexamined.

If there is a very limited idea, it may be due to the fact that the designer has made a fundamental assumption without realizing it. For example, one function typically encountered in attaching is to move in closely. If moving by hand is the only way, then the implicit assumption is that the part will be assembled manually. This may not be a bad assumption, except that it may not have a universal validity. The function may be indicated as how and not what. If the function is to roll on the rail, then the general concept of move has been

	Option 1	Option 2	Option 3	Option 4
Vegetable Picking Device		Triangular plow	Tubular grabber	Mechanical picker
Vegetable Placing Device	Conveyor belt	Rake	Rotating mover	Force from vegetable accumulation
Dirt Sifting Device	Square mesh	Water from well	Slits in plow or carrier	
Packaging Device				
Method of Transportation		Track system	Sled	
Power Source	Hand pushed	Horse drawn	Wind blown	Pedal driven

Figure 7.2 Mechanical vegetable collection system.

121

specifically indicated to the movement along the rail. Designers may have a limited idea of other fields. It is generally important to keep the level of abstraction (details) at the same degree.

Step 2: For each function or functions, list the means or methods to be used.

These secondary lists are the individual subsolutions that, when one is combined from each list, form the overall design solution. These subsolutions can also be expressed in rather general terms, but it is probably better if they can be identified as real devices or subcomponents. For example, when you want to lift, you may like to use a ladder, screw, hydraulic piston, or rack/pinion. The list of solutions should contain not only the known components but also any new or unconventional methods or solutions for achieving the specific task. Many brainstorming or lateral thinking concepts can be employed to arrive at the solution. Remember that many possibilities are necessary for each function. Also, when trying to identify artifacts for a given function, only that specific function or subfunction should be the criterion, not the overall function. Those eliminations will be taken care of at a different time in the design process.

Step 3: Draw up the chart containing all the possible subsolutions.

This chart is called the morphological chart. The morphological chart is constructed from the functional list. At first this is simply a grid of empty squares. The left-hand column lists each function identified in the functional diagram. Across each row, it lists all means (methods) for achieving the function. In the left-hand column, each method is entered in a square. As many methods as possible for each function should be entered. Once finished, the morphological chart contains the complete range of all different theoretically possible solution forms for the product. This complete range of solutions consists of the combinations made up by selecting one subsolution at a time from each row. The total number of combinations is rather large. The list of methods should be kept to about five.

Step 4: Identify feasible combinations.

A list of concepts will be generated from the previous step for each function. The next step is to combine the individual concepts into complete conceptual designs. The method is to select one concept for each function and combine those selected into a single design. Some of these combinations, probably a small number, will be existing solutions; some new solutions; and some (usually a great number) impossible solutions. It is necessary to identify potential solutions for further consideration. Evaluating these solutions is be the subject of Chapter 8.

7.6   DEVELOPING CONCEPTS—SAMPLES

7.6.1   Mechanical Vent

Consider the case involving a design to automate the opening and closing of air conditioning vents in a centralized location in the house (see Chapter 3). Design teams have developed a function analysis wherein the functions are as follows:

1. Select vent
2. Send signal
3. Receive signal
4. Convert signal
5. Open/close vent

Figure 7.3 shows a morphological chart for the problem. The five functions are entered in the left-hand column of the chart. Each row represents a specific function. For example, row 5 is the function "open/close vent", and we need to find means to achieve the function of opening/closing the vent. The chart shows five different means or methods to achieve the function using gears, belt, electric field, cable, or impact plate. The methods are entered using words and diagrams in each of the five squares following "open/close vent."

The chart thus contains the five functions and 5, 5, 5, 6, 5 ways of achieving each function (see Figure 7.3). Thus, theoretically there are 3750 possible different open/close vent machines, combining the various means. However, many of them are not viable. At least five or six of such combinations will be identified for building the system.

7.6.2 Wheelchair Retrieval Unit

A design team is required to design a retrieval unit for wheelchairs to assist nurses performing walking activities with patients. In most events a single nurse is in charge of assisting patients during walking exercise. However, the nurse cannot assist the patient and drag the wheelchair. The design team developed a function analysis for the unit as follows:

1. Align wheelchair to patient and nurse
2. Move wheelchair
3. Steer wheelchair
4. Stop wheelchair

Figure 7.4 shows the morphological chart for the unit. It consists of six different options for each of the four functions, thus you will have 6^4 possible solutions. Again, five or six different viable solutions will be selected out of these solutions. For example, one possible combination is 2, 2, 3, 1 (i.e., rail, track, rail, reverse power).

7.6.3 Automatic Can Crusher

The design team lists all the functions that must be accomplished. Then team members generate a matrix that shows the functions in the right column, and they point out different ways to achieve these functions. Creativity should be exercised during this activity. The different possibilities may be listed in text or in sketches. Both methods are exercised in this situation. The available power sources are as follows:

- Hydraulic
- Magnetic
- Gravity
- Thermal

	Option 1	Option 2	Option 3	Option 4	Option 5	Option 6
Select Vent	Remote switchboard	Handheld remote	1 2-way switch in room	2 1-way switches in room	Dial switch	
Send Signal	Wiring	Transmitter (rf)	Walk to room	Cable/pulley	Hoses (air/hydraulic)	
Receive Signal	Radio receiver	Electrical power device	Lever connected to pulley/cable	Hand lever	Piston (pneumatic/hydraulic)	Hand
Convert Signal	Air compressor	Hydraulic motor	Electric motor	Electromagnet	Combustion	
Open/Close Vent	Gears	Belt	Electric field	Cable	Impact plate	

Figure 7.3 Morphological chart of mechanical vent machine.

124

Figure 7.4   Wheelchair retrieval unit.

- Sound
- Pneumatic
- Solar
- Electric
- Combustion

Several designs can be generated from the morphological chart. Several concepts are sketched in Figure 7.5.

7.7 PROBLEMS

7.7.1 Team Activities

1. Develop a morphological chart for the following statement:
   If you drive in the state of Florida, you may notice some traffic congestion due to the trees being trimmed on state roads and freeways. Assume that the Florida Department of Transportation is the client. Usually, branches 6 inches or less in diameter are trimmed. The allowable horizontal distance from the edge of the road to the branches is 6 ft. The material removed from the trees must be collected and removed from the roadside. To reduce the cost of trimming, a maximum of only two workers can be assigned for each machine. The overall cost, which includes equipment, labor, etc., needs to be reduced by at least 25% from present cost. The state claims that the demand for your machines will follow the price reduction (i.e., If you are able to reduce the cost by 40%, the demand will increase by 40%.) Allowable working hours depend on daylight and weather conditions.
2. Compare (1) the systematic design process, in which functions are created and, based on these functions, a morphological chart is created and thus a large number of possible solutions; with (2) a design process, in which the problem is identified and then a brainstorming session is started to generate different solutions. Which of these methods would you utilize, and why? In the event that a certain aspect of the design needs modification, which of the two methods would be easier to apply toward that modification. Why?
3. Compare and contrast group creativity and individual creativity.
4. Discuss the statement that design is a social activity. How would that statement fit in the systematic process?

7.7.2 Individual Activities

1. What is the difference between creativity and innovation?
2. What is a morphological chart?
3. Without a creative objective tree, you would not have a creative morphological chart. Discuss.
4. How would you increase your creativity level? Categorize yourself (analytical, organizational, social, intuitive). You can perform the survey in the lab exercise to find out (see Lab 1).
5. What is the Herrmann model of the brain?

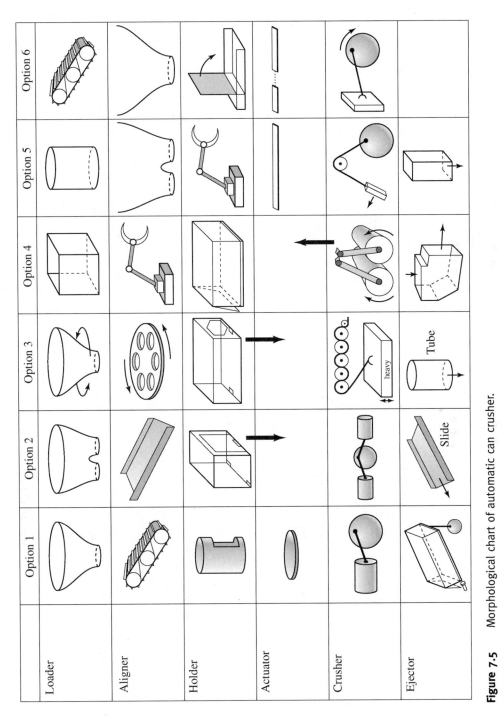

Figure 7-5 Morphological chart of automatic can crusher.

	Option 7	Option 8	Option 9	Option 10		
Loader						
Aligner						
Holder						
Actuator						
Crusher						
Ejector						

Figure 7-5 Morphological chart of automatic can crusher (continued).

7.8 SELECTED BIBLIOGRAPHY

1. Amabile, T. M., and Conti, R. "Changes in the Work Environment for Creativity During Downsizing," *Academy of Management Journal*, Vol. 42, pp. 630–640, 1999.
2. Crandall, R. *Break-Out Creativity*. Corte Madera, CA: Select Press, 1998.
3. Cross, N. *Engineering Design Methods*: *Strategies for Product Design*. New York: Wiley, 1994.
4. Dhillon, B. S. *Engineering Design*: *A Modern Approach*. Toronto: Irwin, 1995.
5. Dunnette, M. D., Campbell, J., and Jaastad, K. "The Effect of Group Participation on Brainstorming Effectiveness for Two Industrial Samples," *Journal of Applied Psychology*, Vol. 47, pp. 30–37, 1963.
6. Dym, C. L. *Engineering Design*: *A Synthesis of Views*. Cambridge, UK: Cambridge University Press, 1994.
7. Gallupe, R. B., Bastianutti, L. M., and Cooper, W. H. "Unblocking BrainStorms," *Journal of Applied Psychology*, Vol. 76, pp. 137–142, 1991.
8. Gardner, H. "Creative Lives and Creative Works: A Synthetic Approach." *In The Nature of Creativity*, J. R. Sternberg (editor), pp. 298–321. Cambridge, UK: Cambridge University Press, 1988.
9. Goldenberg, J., and Mazursky, D. "First We Throw Dust in the Air Then We Claim We Can't See: Navigating in the Creativity Storm," *Creativity and Innovation Management*, Vol. 9, pp. 131–143, 2000.
10. Hennessey, B. A., and Amabile, T. M. "The Conditions of Creativity." In *The Nature of Creativity*, J. R. Sternberg (editor), pp. 11–38. Cambridge, UK: Cambridge University Press, 1988.
11. Herrmann, N. *The Creative Brain*. Lake Lura, NC: Brain Books, 1990.
12. Hill, H. W. "Group Versus Individual Performance: Are N + 1 Heads Better Than One?" *Psychological Bulletin*, Vol. 91, pp. 517–539, 1982.
13. Isaksen, S. G., and Treffinger, D. J. *Creative Problem Solving*: *The Basic Course*. Buffalo, NY: Bearly Limited, 1985.
14. Kemper, J. D. *Engineers and Their Profession*. New York: Oxford University Press, 1990.
15. Kolb, J. "Leadership of Creative Teams," *Journal of Creative Behavior*, Vol. 26, pp. 1–9, 1992.
16. Lumsdaine, E. M., Lumsdaine, M., and Shelnutt, J. W. *Creative Problem Solving and Engineering Design*. New York: McGraw-Hill, 1999.
17. McCabe, M. P. "Influnce of Creativity on Academic Performance," *Journal of Creative Behavior*, Vol. 25, p. 2, 1991.
18. McKim, R. H. *Experiences in Visual Thinking*. Boston, MA: PWS Publishing Co., 1980.
19. Mumford, M. D., and Gustafson, S. B. "Creativity Syndrome: Integration, Application and Innovation," *Psychological Bulletin*, Vol. 103, pp. 27–43, 1988.
20. Nickerson, R. S. "Enhancing Creativity," *In The Nature of Creativity*, J. R. Sternberg (editor), pp. 392–430. Cambridge UK: Cambridge, University Press, 1988.
21. Pahl, G., and Beitz, W. *Engineering Design*: *A Systematic Approach*. Springer-Verlag, 1996.
22. Ulman, D. G. *The Mechanical Design Process*. New York: McGraw-Hill, 1992.
23. Ulrich, K. T., and Eppinger, S. D. *Product Design and Development*. New York: McGraw-Hill, 1995.
24. Weisberg, R. W. *Creativity. Genius and Other Myths*. New York: Freeman, 1986.
25. White, D. "Stimulating Innovative Thinking," *Research-Technology Management*, Vol. 39, pp. 31–35, 1996.

An engineer or design team will determine a product's functional alternatives from its morphological chart. This will assist the design team in finding the best design for the client's need. Having chosen a set of alternatives, the design team can start sketching them as devices. These vents were the result of numerous design sketches. (Copyright Sami Sarkis/Getty Images/PhotoDisc.)

8.1 OBJECTIVES

By the end of this chapter, you will be able to use the Pugh method to evaluate the different concepts that were generated in previous design step.

In the previous chapter, we developed a technique for generating different concepts that satisfy the need statement. As was demonstrated, the number of possibilities that may satisfy the need is a function of the number of entries (different ways) that fulfill a function. The larger the number of entries, the larger the number of concepts generated. Obviously, not all combinations through the morphological chart will lead to a feasible product. The goal of this chapter is to provide a methodology for evaluating the different devices. By the end of the evaluation process, a product that has the highest potential for becoming the quality product will be identified through the many concepts that were generated by the morphological chart.

The difficulty in concept evaluation resides in that the design team members must choose a concept to develop based on limited knowledge and data. Concepts at this point are still abstract, have little details, and cannot be measured. The essential question then becomes, "Should time be spent refining and adding minute details to all concepts and then measuring their outputs?" or "Would developing a mechanism to point out one concept that will most likely become the quality product prove to be a more efficient option?" The screening of concepts can be accomplished by using the following levels:

1. The concepts are directly measured with respect to a fixed set of limits (specifications). If a concept doesn't satisfy the specification requirement, it will be dropped. Although certain functional presentations are on the morphological chart, they will be eliminated from assembly if they do not meet the specification criteria established earlier. For example, using sonar to align the wheelchair may be eliminated in favor of another functional presentation that conforms to the projects cost limitations.

2. The design team uses their experience (engineering sense) to eliminate some concepts because the technology is not yet available to facilitate the concept. Or, based on their judgment, the product is not feasible. Why use a parachute to stop the wheelchair if another function that is more sound (from an engineering standpoint) could perform the same function?

3. The concepts are evaluated with respect to each other using measures defined by the criteria (specifications/wishes). This screening method was developed by Stuart Pugh in the 1980s, and is often called Pugh concept selection. The purpose of this stage is to narrow the number of concepts quickly. This evaluation matrix will be presented in more detail in this chapter.

8.2 SKETCH ASSEMBLY OF ALTERNATIVES

Using the different functional alternatives presented in the morphological chart, your engineering sense, and a specification table, choose a set of five to six alternatives. At this stage you have started the convergent part of the design process that will lead toward a best design for the need at hand.

Having chosen the set of alternatives, you need to sketch them as devices, not as a set of functions. The conceptual sketches are not detailed. The sketches serve as a first presentation of the prospective design. Also, sketching reveals information about details that may be needed later, which will also help in evaluating the concepts. Assembly of the functional mechanisms will help the visual conceptualization of the alternatives and will provide a presentation tool to the customer. Figure 8.1 shows six different alternatives generated for the mechanical vent.

Sketches are the artistic expression of the design. Visualizations of the creative ideas are presented in sketches. Students need to master the craft of sketching.

Figure 8.1 Conceptual sketches of the mechanical vent.

8.3   DECISION MATRIX

The decision matrix method, or Pugh method, is based on comparing alternatives with the wishes list in the specification table. An alternative should meet the customer demands; otherwise it will be dropped in the initial screening. Also, the concept should seem feasible to the engineering group or it will be dropped by the second stage of screening. The concepts that pass the initial two stages need to be evaluated with respect to each other, using fixed criteria (specification table). The Pugh evaluation method tests the completeness and understanding of requirements, rapidly identifies the strongest alternative, and helps foster new alternatives. This method is most effective if each member of the design team performs it independently, and the individual results are then compared. The results of the comparison lead to a repetition of the technique, with the iteration continued until the team is satisfied with the results. The Pugh evaluation method is performed using the following steps:

1. *Choose the comparison criteria.* If all alternatives fulfill the demands on the same level, then the criteria should be the wishes in the specification table. Remember, the wishes were organized in an ordered fashion before they were used in the house of quality. The order is based on the level of importance according to the design team. Different weighing schemes could be used in the ordering of wishes, such as
   a. Absolute factor, where each wish is evaluated individually on a scale from 0 to 10. Other wishes will not interfere with the weigh factor used.
   b. Relative scale, where the sum of the wishes is assigned a scale from 0 to 100. Each wish is assigned a weighing factor that corresponds to its importance relative to the other wishes. However, when the weighing factors are added, they should equal 100.
   In this book, the choice of the evaluation scheme is given to the design team. The weighting factor method is based on engineering team's assessment of how important the attribute may be to the final product. If the alternatives contain differences in the levels at which demands may be fulfilled, then the demands can be entered in the comparison criteria.

2. *Select the alternatives to be compared.* From the morphological chart, different alternatives will be generated. Some of these alternatives will be dropped because they do not satisfy the customer demands or they are not feasible. The rest of the alternatives are possible candidates. However, by using the initial screening stages, a few alternatives will be left for the final stage. Not all feasible alternatives will be allowed to enter the final stage. Up to six will be allowed.

3. *Generate scores.* After careful consideration, the design team chooses a concept to become the benchmark or datum against which all other concepts are rated. The reference is generally either
   a. An industry standard, which can be commercially available product or an earlier generation of the product.
   b. An obvious solution to the problem.

c. The most favorable (measured according to a vote by the design team) concepts from the alternatives under consideration.

d. A combination of subsystems, which have been combined to represent the best features of different products.

For each comparison, the concept being evaluated is judged to be better than, about the same as, or worse than the datum. If it is better than the datum, the concept is given a [+] score. If it is judged to be about the same as the datum, the concept is given [0]. If the concept does not meet the criterion as well as the datum does, it is given a [−].

4. *Compute the total score.* After a concept is compared with the datum for each criterion, four scores are generated: the number of plus scores, the number of minus scores, the overall total, and the weighted total. The overall total is the difference between the number of plus scores and the number of minus scores. The weighted total is the sum of each score multiplied by the weighing factor. The scores can be interpreted in a number of ways:

a. If a concept or a group of similar concepts has a good overall total score, or a high + total score, it is important to notice what strengths it exhibits. In other words, notice which criteria it meets better than the datum. Likewise, the grouping of − scores will show which requirements are difficult to meet.

b. If most concepts get the same score on a certain criterion, examine that criterion closely. It may be necessary to develop more knowledge in the area of the criterion to generate better concepts. Or, it may be that the criterion is ambiguous, or it may be interpreted differently by different members of the design team.

c. To learn even more, redo the comparison, with the highest scoring concept used as the new datum. This iteration should be redone until the clearly best concept or concepts emerge. *This process becomes a must when the datum is one of the alternatives under consideration.*

After each team member has completed this procedure, the entire team should compare their individual results. Then a team evaluation needs to be conducted.

8.4 CONCEPTS EVALUATION: MACHINE SHOP KIT

A team of design students was asked to design a steam-powered machine shop kit that can be used to (1) develop hands-on skills in using machine shop tools for freshman engineering students, and (2) demonstrate the conversion of thermal energy into work, thus becoming a demonstration tool for introduction to thermal science class. Stirling engine kits are being used in many engineering schools. The new kit must compete with the Stirling engine kit in its educational value and its cost. A Stirling engine kit is a kit containing the disassembled parts that compose a

Figure 8.2 Function analysis of machine shop kit.

Stirling engine; a few of the parts require students to use the equipment in the workshop. Students will be trained on the workshop equipment, their assemblies, and the mechanism on which the machine works!

Figure 8.2 shows the functional analysis that the design team developed. The energy track of the functions consists of receive energy, channel energy, store energy, transmit energy, and utilize energy. Figure 8.3 shows the house of quality of the problem. Figure 8.4 shows the morphological chart developed by the design team. Seven different conceptual designs were developed in this exercise. Figures 8.5 through 8.10 show the six different concepts.

Using the evaluation chart, each design was rated against a datum, the Stirling engine, which is currently used in the machine shop. The rating criteria were based on the design specification. A positive sign indicates that a design has a better rating than the datum. A negative sign, on the other hand, denotes that the design is less desirable than the datum in that particular category. If it so happens that the design is neither better nor worse at meeting a specification than the datum, then the value of zero is assigned in the chart. After carefully evaluating each design, the total number of positives and negatives was calculated. Furthermore, the weighted total was determined by multiplying each positive or negative by the specification weight and then adding them. Upon completing the design evaluations, it was concluded that Concept VI best met the requirements and was therefore chosen as our project design. Figure 8.11 shows the evaluation table.

	Assembly time (<20 hrs)	# of parts (10–15)	Pollution factor (<9 ppm)	Injury probability (<0.1%)	Retail price (<$135)	Replace parts cost (<5–10% total cost)	Efficient (>40%)	Vibrations (<2/sec)	Noise (<60 dB)	Dimensions (1' × 6" × 6")	Weight (<20 lbs)	Users find visually pleasing (75%)
Assembly												
Easy to assemble	9	3		1								
Easy to disassemble	9	3		1								
Moderate assembly time	9	3										
Interesting to build	1											
Not too many parts	9	9			3	1		3	1	1	3	
Safety												
Low pollution			9				1					
No flying debris				9								
No sharp edges				9								
Costs												
Retails for less than the competition					9	1						
Low replacement part costs					3	9						
Inexpensive materials					9	3					3	1
Performance												
Convert energy efficiently			3		1		9	3				
Low vibration		1					3	9	3		1	
Runs off small amount of energy							3	1				
Low noise							1	3	9			
Physical Requirements												
Portable										9	9	
Strong material				3				1			3	
Corrosion proof				1								3
Lightweight		3		1				1		3	9	
Visually appealing												9

Figure 8.3   House of quality for machine shop kit.

	Option 1	Option 2	Option 3	Option 4	Option 5	Option 6	Option 7
Receive	Open cylinder	Spring	Closed cylinder				
Channel	Funnel	Linkage	Shaft	Gear	Tube	Piston	
Store	Flywheel	Piston	Capacitor	Propeller	Shaft	Tube	
Transmit	Shaft	Belt	Gear	Steam wheel			
Use	Wheel & axle	Rod	Propeller	Linkage	Gear	Flywheel	Pulley

Figure 8.4 Morphological chart for machine shop kit.

Figure 8.5 Concept I of machine shop kit: A tank full of water is heated to produce steam. The steam will travel through the tube and push the piston, which will turn the attached flywheel.

Figure 8.6 Concept II of machine shop kit: A flame is used to heat a piston, which will be pushed out to turn a gear. At the same time that the piston is pushed out, another piston is being pushed up, which will push the hot air piston back to its original position.

Figure 8.7   Concept III of machine shop kit: This design incorporates a system of pistons. The first piston is pushed by the pressure from heated air. It, in turn, compresses a medium of oil, which causes the final piston to be pushed.

Figure 8.8   Concept IV of machine shop kit: Hot air is funneled to turn a propeller system. The propeller is connected to a central rod, which has a gear attached to it. The rotation of blades will cause the attached gear to rotate, which turns the other gear.

Figure 8.9   Concept V of machine shop kit: The hot air is channeled, which causes the propeller to rotate, which spins a flywheel. The flywheel is connected to a second flywheel by a connector link. Therefore, as the first flywheel turns, the second flywheel will also turn.

Figure 8.10   Concept VI of machine shop kit: Two metal tanks filled with water are heated with an alcohol burner. The heated water then generates steam that travels through a nylon tube to a steam tube. The steam tube is connected to two "steam wheels," which have holes drilled in them at 90 angles. The escaping steam will create rotation, which will turn the axles that turn the wheels and move the car.

Evaluation Chart	Sketch 1	Sketch 2	Sketch 3	Sketch 4	Sketch 5	Sketch 6	D
Easy to assemble 7	0	0	0	+	0	+	A
Easy to disassemble 7	0	0	0	+	+	+	T
Safe for operator 10	0	0	0	0	0	0	U
Low vibration 5	+	−	+		0	0	M
Portable 4	−		0	0	0	0	+
No sharp edges 6	+		0	+	−	−	0
Retails for less than competition 9	+	+	+	+	+	+	
Convert energy efficiently 10	−		0	0	0	0	0
No flying debris 8		0	0	0	0	0	0
Low pollution 3		0	0	0	0	0	0
Low replacement part cost 7	+		0	0	+	+	+
Low noise 4		0	+	+	0	0	+
Strong material 6		0	0	0	0	0	−
Low energy dissipation 8	+		0	0	−	0	−
Aesthetically appealing 5	−		0	−		0	0	+
Total +	5	2	4	4	5	7	
Total −	3	1	1	2	1	2	
Overall total	2	1	3	2	4	5	
Weighted total	16	8	19	16	22	29	

Figure 8.11   Evaluation table for the machine shop kit.

8.5   CONCEPTS EVALUATION: AUTOMATIC CAN CRUSHER

From the morphological chart, many possible combinations of devices can be obtained, but not all of these processes are positive or able to be accomplished effectively. In this stage, the design team begins to limit the possible designs based on specifications, manufacturability, cost, and other factors. The surviving designs are then evaluated against each other. The concepts are shown in Figures 8.12 through 8.18. A decision matrix is constructed to evaluate the designs based on the wishes obtained from the specification table. Specifications are weighted by importance, with 1 being the lowest and 10 being the highest. The concepts are evaluated and one design is chosen by the design team to be the best possible design. This design is considered the datum by which all of the other designs are compared. Then, all of the other designs are evaluated against the datum and receive a + if they exceed the datum in the wish, a − if they are below the datum in the wish, and 0 if they are equal to the datum in the wish. The results are tallied and scores are evaluated; then the best design is placed as the datum, and the evaluation is done again. The highest scoring design is the chosen design to be built.

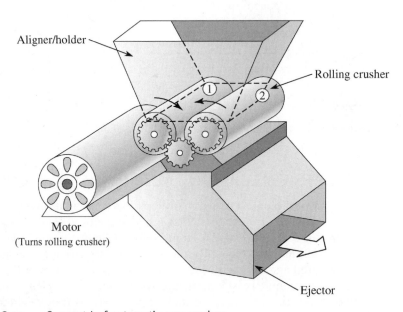

Figure 8.12 Concept I of automatic can crusher.

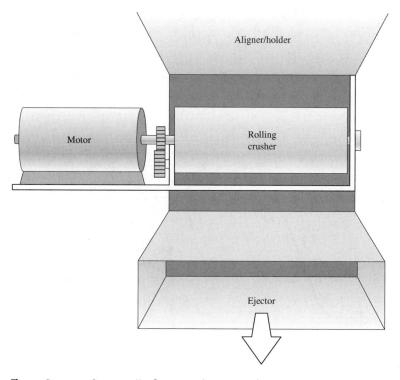

Figure 8.13   Concept II of automatic can crusher.

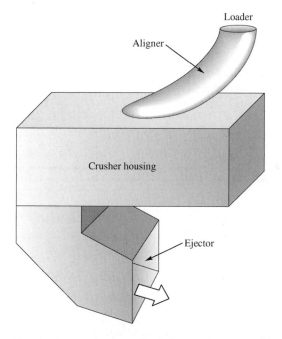

Figure 8.14   Concept III of automatic can crusher.

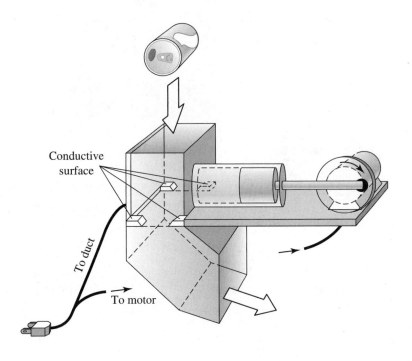

Conductive
surface

To duct

To motor

Can completes
broken circuit

Power from
motor

Connection
to inside
surface

Split

Power to motor

Figure 8.15   Concept IV of automatic can crusher.

Loader

Aligner

Holder

Actuators

Piston

Flywheel   Motor

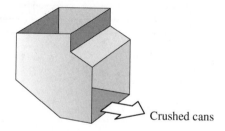

Crushed cans

Figure 8.16   Concept V of automatic can crusher.

Critertion	Wt	Concepts				
		I	II	III	IV	V
inoperable when opened	10	0	0	0	0	
runs on standard 110-V outlet	7	0	0	0	0	
< $50 retail	9	0	0	0	0	
enough force to crush glass and plastic	3	0	0	0	0	D
large storage of crushed cans	5	0	0	0	0	A
easily accessible kill switch	3	0	0	0	0	T
no sharp corners	10	0	0	0	0	U
easy to start	8	0	0	0	0	M
stops easily and immediately	9	0	0	0	0	
can stop in mid operation	8	0	0	0	0	
drains residual fluid from machine	10	0	0	0	0	
low vibration	8	0	0	0	0	
low peripheral force	8	0	0	0	0	
shock absorption	8	0	0	0	0	
high-efficiency engine	8	0	0	0	0	
high material strength	9	0	0	0	0	
small force to depress switches	9	0	0	0	0	
safety stickers	9	0	0	0	0	
reset button	9	0	0	0	0	
no flying debris	9	0	0	0	0	
operating steps stickers	9	0	0	0	0	
low loading height	7	0	0	0	0	
easily accessible interior	7	0	0	0	0	
crushing mechanism inaccessible	10	−	0	−	0	
crush many cans a minute	9	+	+	0	0	
utilize gravity in design	4	0	0	0	0	
low noise output	5	0	0	0	0	
starts up immediately	5	0	0	0	0	
easy access to clear jams	6	−	−	0	0	
large capacity loader	5	+	+	0	0	
long running capability	7	0	0	0	0	
parts easy to acquire	8	−	−	0	0	
internal parts safe from liquid damage	9	0	0	0	0	
low operating cost	5	0	0	0	0	
easy cleaning	6	−	−	0	0	
easy to disassemble	6	0	0	0	0	
crushes along major axis	2	0	0	0	−	
no exhaust	2	0	0	0	0	
Total +		2	2	0	0	
Total −		4	3	1	1	
Overall total		−2	−1	−1	−1	
Weighted total		−16	−6	−10	−2	

Figure 8.17 Evaluation of designs: V is datum.

Critertion	Wt			Concepts		
		I	II	III	IV	V
inoperable when opened	10	0	0	0	0	
runs on standard 110-V outlet	7	0	0	0	0	
< $50 retail	9	0	0	0	0	
enough force to crush glass and plastic	3	0	0	0	0	D
large storage of crushed cans	5	0	0	0	0	A
easily accessible kill switch	3	0	0	0	0	T
no sharp corners	10	0	0	0	0	U
easy to start	8	0	0	0	0	M
stops easily and immediately	9	0	0	0	0	
can stop in mid operation	8	0	0	0	0	
drains residual fluid from machine	10	0	0	0	0	
low vibration	8	0	0	0	0	
low peripheral force	8	0	0	0	0	
shock absorption	8	0	0	0	0	
high-efficiency engine	8	0	0	0	0	
high material strength	9	0	0	0	0	
small force to depress switches	9	0	0	0	0	
safety stickers	9	0	0	0	0	
reset button	9	0	0	0	0	
no flying debris	9	0	0	0	0	
operating steps stickers	9	0	0	0	0	
low loading height	7	0	0	0	0	
easily accessible interior	7	0	0	0	0	
crushing mechanism inaccessible	10	−	0	−	0	
crush many cans a minute	9	+	+	0	0	
utilize gravity in design	4	0	0	0	0	
low noise output	5	0	0	0	0	
starts up immediately	5	0	0	0	0	
easy access to clear jams	6	−	−	0	0	
large capacity loader	5	+	+	0	0	
long running capability	7	0	0	0	0	
parts easy to acquire	8	−	−	0	0	
internal parts safe from liquid damage	9	0	0	0	0	
low operating cost	5	0	0	0	0	
easy cleaning	6	−	−	0	0	
easy to disassemble	6	0	0	0	0	
crushes along major axis	2	0	0	0	+	
no exhaust	2	0	0	0	0	
Total +		2	2	0	1	
Total −		4	3	1	0	
Overall total		−2	−1	−1	1	
Weighted total		−16	−6	−10	2	

Figure 8.18 Evaluation of designs: IV is datum.

8.6   PROBLEMS

8.6.1   Team Activities

1. In order to enhance your engineering sketch ability, consider the paper stapler. Disassemble the stapler to its basic components. Sketch these components and sketch an assembled stapler.
2. Why would a designer place items (demos of function) in the morphological chart if he or she knows that they would not be feasible?
3. Define engineering sense. If someone asks you to demonstrate a speed of 10 cm/sec, how would you do so using your engineering sense?
4. Why would you need to evaluate alternatives if your gut feeling is pointing you toward one of these alternatives?

8.6.2   Individual Activities

1. Discuss the following: "Using different alternatives for each function allows design engineers to substitute that particular element in case of a problem rather than changing the whole design."
2. What difference does it make if you use the absolute or relative scale in the weighing function? Which one would you recommend and why?
3. Why would you evaluate the different alternatives individually rather than as a team?
4. In all examples presented in this chapter, the design team has the opportunity to establish the evaluation criteria (which may thus be biased by the design team). What would you use as mechanism to establish the evaluation criteria if you are the design manager at Ford Motor Company? How would you conduct the evaluation?
5. How did the house of quality contribute to the evaluation mechanism?
6. In the event that the mechanism that satisfies a function, or its presentation, is found to have difficulty in attaching to another function(s), what would you do and why?
7. Name situations in which you have to perform the evaluation chart more than once.
8. Use the Pugh method to evaluate this course. Use your physics course as a datum.

8.7   SELECTED BIBLIOGRAPHY

1. Ambrose, S. A., and Amon, C. H. "Systematic Design of a First-Year Mechanical Engineering Course at Carnegie Mellon University," *Journal of Engineering Education*, pp. 173–181, 1997.
2. Burghardt, M. D. *Introduction to Engineering Design and Problem Solving.* New York: McGraw-Hill, 1999.
3. Cross, N., Christian, H., and Dorst, K. *Analysing Design Activity.* New York: Wiley, 1996.

4. Eekels, J., and Rooznburg, N. F. M. "A Methodological Comparison of Structures of Scientific Research and Engineering Design: Their Similarities and Differences," *Design Studies*, Vol. 12, No. 4, pp. 197–203, 1991.

5. Pahl, G., and Beitz, W. *Engineering Design: A Systematic Approach.* New York: Springer-Verlag, 1996.

6. Pugh, S. *Total Design.* Reading, MA: Addison-Wesley, 1990.

7. Suh, N. P. *The Principles of Design.* New York: Oxford University Press, 1990.

8. Ulman, D. G. *The Mechanical Design Process.* New York: McGraw-Hill, 1992.

9. Ulrich, K. T., and Eppinger, S. D. *Product Design and Development.* New York: McGraw-Hill, 1995.

10. Vidosic, J. P. *Elements of Engineering Design.* New York: The Ronald Press Co., 1969.

11. Walton, J. *Engineering Design: From Art to Practice.* New York: West Publishing Company, 1991.

9 | Embodiment Design

At this stage of the design process, the design team starts the embodiment design stage by presenting information through concept sketches. These sketches are then fine tuned into product drawings.

9.1   OBJECTIVES

By the end of this chapter, you should be able to

1.   Discuss the different types of presentations of a product.
2.   Discuss the difference between prototype and mock-up.
3.   Construct a bill of material.
4.   The terminology for design for "X"

In the previous two chapters, different devices that perform the requirements were derived by using the morphological chart. The concepts were sketched, not in great detail, but in a way that presents the major functions that compose the device. These concepts were then evaluated based on three different stages. The first stage involves subjecting the concept to the specification table. If it passes the demands, then a feasibility test is conducted. If the concept passes the first two stages, then it will be subjected to relative evaluation based on the Pugh method. At this stage of the design process, the design team starts what can be called the embodiment design stage. The word *embody* means "to give something a tangible expression".

This chapter discusses presentation of the concept(s) that pass the evaluation criteria. The idea of presenting information through drawings was introduced when different concepts were generated from the morphological chart. The sketches that present the concepts are not detailed: They simply provide enough information to demonstrate the combination of the different function mechanisms that produce the concept. Figure 9.1 shows conceptual sketches of a four yard leaf collector.

It should be reemphasized that the design process is an iterative process. This means that if additional refinements are found necessary, after the skeleton sketches or while in the evaluation stage, the designer should iterate back to the concept phase and possibly generate new concepts. The drawback of this process is that it takes time, but that time should be initially implanted in the scheduling of events. The use of functional structure and morphological charts makes it easy on the designer to replace components in the design rather than start over if a problem arises. The objective tree and the house of quality provide the evaluation criteria. These steps have well-defined boundaries thus facilitate the iterative process.

In the world of engineering, information needs to be passed on through drawings. The analysis, manufacturing, and assembly of the product is based on drawings. The following section discusses the type of drawings that are required by the design team.

Figure 9.1   Conceptual sketches of yard leaf collector.

9.2 PRODUCT DRAWINGS

In the design process, drawings may be used to serve the following objectives:

1. Drawings are the preferred form of data communications in the world of engineering.
2. Drawings are used to communicate ideas between the designers and between the designers and manufacturing personnel.
3. Drawings simulate the operation of the product.
4. Drawings are used to check for the completeness of the product.

There are different types of drawings that can be used at different levels:

1. *Sketches (layout drawing):* Sketches are generated during the conceptualization stage of the design process. Different mechanisms of each function or subfunction are first sketched. These sketches are then used to build devices. These sketches present the completeness of the device. They are working documents that support the development of the major components and their relations. They are made to scale, but only the important dimensions are shown in the layout. Tolerances are usually not shown unless they are critical. This form of drawing is needed in the analysis. When an individual component is analyzed, a sketch presents the relation of this part to its surroundings.

2. *Detail drawings:* As the product evolves, a detail of individual components develops. A typical detail drawing is shown in Figure 9.2. The important characteristics of a detail drawing are as follows:
   a. All dimensions must be tolerances.
   b. Materials and manufacturing detail must be clear and presented in specific language.

These drawings are made using a computer-aided design (CAD) system, such as Pro/E.

Figure 9.2 Typical detail drawing.

No	Description	QTY.
1	Frame	1
2	Bearing	8
3	Wheel	4
4	Axle	4

Drawing Name: Skate Assembly		
Drawn by: Nicholas Toro Checked by: Tikishia Jackson		
Scale	Date	Draw. No.
0.5	1/25/2002	01

Figure 9.3   Typical assembly drawing.

3. *Assembly drawings:* The goal of the assembly drawing is to show how components fit together. Figure 9.3 shows a typical assembly drawing. An assembly drawing has the following characteristics:
   a. Each component is identified with a number referenced to a list of all components.
   b. Necessary, detailed views are included to explain information that is not clear in other views.

9.3   BILL OF MATERIAL

The bill of material is an index of the parts that were used in the product. A typical bill of material is shown in Figure 9.4.

The following information should be included in a bill of material:

1. *The item number:* This is a key to the components on the assembly drawing.
2. *The part number:* This is a number used throughout the purchasing, manufacturing, and assembly system to identify the component. The item number is a specific index to the assembly drawing; the part number is an index to the company system.

Item	Part	Quantity	Name	Material	Source
1	G-9042-1	1	Governer body	Cast aluminum	Lowe's
2	G-9138-3	1	Governer flange	Cast aluminum	Lowe's
9	X-1784	4	Governer bolt	Plated steel	Fred's Fine Foundry

Figure 9.4   Bill of materials.

3. *The quantity needed in the assembly.*
4. *The name and description of the component.*
5. *The material from which the component is made.*
6. *The source of the component.*
7. *The cost of the individual component:* This part will be kept for the design team.

9.4   PROTOTYPE

The experimental phase of the engineering design requires the transformation of paper drawings into hardware, which is constructed and tested to verify the concept's workability.

Four construction techniques are available for designers:

1. *Mock-up:* This is generally constructed to scale from plastics, wood, cardboard, and so forth to give the designer a feel for his or her design. It is often used to check the clearance, assembly techniques, manufacturing considerations, and appearance. This is the least expensive technique, is relatively easy to produce, and can be used as a tool in selling the idea to clients or management. A solid model in the CAD system can often replace the mock-up. Mock-ups are usually referred to as a proof of concept prototype.

2. *Model:* The model relates the physical behavior of the system through mathematical similitude. The modeling is usually referred to as a proof of product prototype. There are different types of models used to predict the behavior of a real system:
   a. *True* is an exact geometric reproduction of the real system, built to scale and satisfying all restrictions imposed by the design parameters.
   b. *Adequate* is constructed to test specific characteristics of the design and is not intended to yield information concerning the total design.
   c. *Distorted* purposely violates one or more design conditions. This violation is often required when it is difficult or impossible to satisfy the specific conditions due to time, material, or physical characteristics, and it is felt that reliable information can be obtained through the distortion.

3. *Prototype:* This is the most expensive technique and the one that produces the greatest amount of useful information. The prototype is a constructed, full-scale working physical system. Prototypes can be comprehensive or focused. A comprehensive prototype corresponds to a full-scale, fully

operational version of the whole product. An example of a comprehensive prototype is the beta prototype, which is given to customers in order to identify any remaining design flaws, before committing to production. In contrast, focused prototypes correspond to one or a few of the product elements. Examples of focused prototypes include foam models used to explore different forms of a product.

4. *Virtual prototyping:* Computer-aided design and computer-aided engineering (CAE) software suppliers have fulfilled the goals of virtual prototyping through the delivery of 3-D feature-based modeling capabilities. Solid modeling has enabled the use of geometry to visualize product models quickly and to detect gross interference problems. One of the unique features in feature-based modeling software is its ability to automate design changes. It provides integrated capabilities for creating detailed solid and sheet metal components, building assemblies, designing weldments, and producing fully documented production drawings and photorealistic renderings. Feature-based modeling is built on combining the commands needed to produce a common feature. For example, a combination of commands to create a hole of prescribed dimensions and relations to keep the parts of the hole maintained is stored in a feature that is given the name *hole.* Parametric modeling allows the presentation of dimensions of the parts in term of parameters. Virtual prototyping can have its biggest payoff in manufacturing and analysis. Using solid models as the base for a finite element analysis tool would reduce prototyping time and save money. In manufacturing, parametric feature-based models will reduce the time needed to produce a feasible prototype.

9.5 ANALYSIS

Once the product is well defined, a complete analysis is conducted. In the analysis process, you need to evaluate the integrity of the design in terms of its safety concerns, choose the material that fits the requirements, and compute the cost involved. Analysis proceeds as follows:

1. Check for the design safety in terms of computing the forces that act on each component. Evaluate the stresses associated with these components. Then list the materials that would satisfy the stress requirements. Students are assumed to have some physical background from their physics classes before this course. It is assumed that the concept of force is known. The material selection in Lab 6 introduces the concept of stress. Students at this stage are also encouraged to seek experts' opinion on the analysis. Software that incorporates the CAD model into analysis mode is available. Students should investigate the availability of such software at their institution.

2. Manufacturability of the components needs to be checked. Having a sketch of a component doesn't mean that the machinist will able to produce the component. You need to provide the machinist with engineering drawings

before the product component can be made. The engineering drawings should be detailed. Use CAD software to produce your drawings.

3. Cost analysis of the components (obtained from a vendor or produced at the machine shop) needs to be conducted. You will be able to vary among components listed in the morphological chart if certain components are cost effective or fit the product better.

4. The aesthetics or the "look" of the product needs to be maintained. Several customer surveys have shown that customers value the look of the product much more highly than its performance.

9.6 DESIGN FOR "X"

Unlike smaller engineering design projects, in an industrial setting, many teams contribute to create new products including non-engineering teams. Several attributes are put into the product to keep up with competition and to produce reliable, marketable, safer, and less costly products in shorter periods of times. In order to integrate these attributes through various activities and maintain cost-effective benefits, a process called "concurrent engineering" has been developed. The use of this approach implies systems theory thinking. Synergistic results become the main goal of its processes.

Designing for such different attributes such as manufacturability, assembly, environment, safety, etc. are usually referred to as a design for "X".

9.6.1 Design for Manufacturing

Design for manufacturing (DFM) is based on minimizing the costs of production including minimizing the time to market while maintaining a high standard of quality for the product. DFM provides guidance in the selection of materials and processes and generates piece part and tooling cost estimates at any stage of product design. You can find several companies on the Internet that sell products and software to facilitate the functioning of a company to perform elements for DFM. This would include:

1. An accurate cost estimator that reviews the cost of parts as they are being designed in a fast and accurate way;

2. A concurrent engineering implementation that provides quantitative cost information which allows the design team to make decisions based on real-time information and to shorten the product development time;

3. Provide supplier negotiations with unbiased details of cost drivers; and

4. Competitive benchmarking that compares the designs with competitors products to determine marketability and target cost.

There are several methods that have been used in DFM to assess in analysis such as process-driven design, group technology, failure mode and effect analysis, value engineering, and the Taguchi method.

9.6.2 Design for Assembly

Design for assembly (DFA) is the study of the ease of assembling various parts and components into a final product. A lower number of parts and an ease of assembly contribute to reducing the overall cost of the product. With DFA, every part has to be checked. It must be determined if it is a necessary part or would it be better integrated into other parts or be replaced by a similar function part that is simpler and costs less. Integrating both design for manufacturing and design for assembly helps contribute to the competitive success of any given product by matching that product's demands to its manufacturability and assembly capabilities.

9.6.3 Design for Environment

Recently, a trend of environmentally conscious products along with the rapid implementation of worldwide environmental legislation have put the responsibility for the end-of-life disposal of products on to the manufacturer. Manufacturers must now implement rules during the design that enforce *design for environment* (DFE). By designing products up-front for environmental and cost efficiency, manufacturers are gaining an edge on slow-to-react competitors who face the same issues unprepared. DFE is concerned with the disassembly of products at the end of their useful life and reveals the associated cost benefits and environmental impacts of a product's design. This quantitative information can then be used to make informed decisions at the earliest concept stages of the product's design.

9.7 EMBODIMENT DESIGN: MECHANICAL VEGETABLE HARVESTING MACHINE

In this example, a design team was asked to design and perform structural analysis on a mechanical vegetable collection and packaging system that is composed mainly of wood. The vegetable harvesting and packaging system must be capable of harvesting the vegetables without damaging a substantial amount of the crop. In addition, the design must not rely on manmade power in order to operate it since it is to be used in the field. During the design process, the design team developed the function analysis shown in Figure 9.5. A morphological chart for the design is shown in Figure 9.6. Three different concepts were generated. Figures 9.7, 9.8, and 9.9 show the three different concepts.

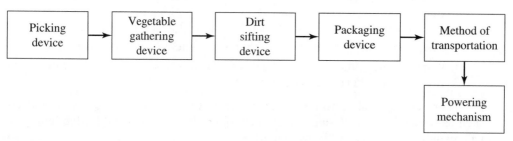

Figure 9.5   Function analysis for vegetable harvesting machine.

	Option 1	Option 2	Option 3	Option 4
Vegetable Picking Device	Conveyor belt	Triangular plow	Tubular grabber	Mechanical picker
Vegetable Placing Device	Square mesh	Rake	Rotating mover	Force from vegetable accumulation
Dirt Sifting Device		Water from well	Slits in plow or carrier	
Packaging Device		Track system	Sled	
Method of Transportation				
Power Source	Hand pushed	Horse drawn	Wind blown	Pedal driven

Figure 9.6   Morphological chart for vegetable harvesting machine.

Figure 9.7   Concept I for vegetable harvesting machine.

Figure 9.8   Concept II for vegetable harvesting machine.

Figure 9.9 Concept III for vegetable harvesting machine.

In the first concept, a horse pulls the yoke moving the vehicle forward. The front wheels turn a belt, which in turns provides torque on the blade shaft. Blades and sifters sort debris from the vegetables. Vegetables are then slung into the carriage compartment where an operator packages product.

The principle on which Concept II's mechanism operates is that vegetables are harvested very close to the soil surface. By taking advantage of a plow, the vegetables will be ripped free from the roots and travel up the plow to the conveyor system. The only piece of metal found on the apparatus is found at the edge of the plow. This edge is reinforced with steel since it will be in direct contact with the soil. The conveyor system is composed of two axles and a conveyor belt; the belt is composed of hemp (the strongest natural plant fiber). The conveyor system was used for sorting out the vegetables from the dirt and small stones, more or less like a sifter. The vegetables will stay on the conveyor system and the soil will fall through the sorter. The rear axle drives the conveyor system via a three-to-one gear ratio from the rear axle to the driving axle of the conveyor system. This allows the conveyor to travel three times the speed of the tangential velocity of the rear axle. Once at the top of the conveyor system, the vegetables are pushed onto a rear-loading table, complete with guides. The rear axle is attached rigidly to the rear axle via a single wooden pin per wheel. The wheel has a carved out section where this pin lies. There is also a wedge that holds the wheel firmly against the pin to ensure a rigid connection.

The conveyor is driven via a friction belt located between the rear axle gears. The belt is arranged in a figure eight to allow the rotation induced upon the belt by the rear axle to be reversed so that the conveyor can travel in the correct direction, towards the back of the machine. The chassis is designed using a single piece of wood. Openings are included in the bottom of the chassis to allow soil or rocks to fall back to the earth. The main portion of the plow is made from the chassis, but is replaceable if severely damaged. The attachments where the horses are fastened are also made from the chassis. All of the wooden pieces are composed of Red Oak.

In Concept III, a horse pulls the device in the y direction. The front wheels turn making the collecting wheel rotate. The tubes cut the vegetables, dragging them against the cart. Once the vegetables are against cart, they are guided inside the pipes to the upper position. There is a hole at the bottom of each pipe where the vegetables exit. At this point, there is a slide that carries the vegetables to the back of the cart.

All three concepts were evaluated for their structural integrity. For the purpose of this presentation, the structural analysis of the second concept will be presented. The analysis was done using Pro/Mechanica software. Pro/Mechanica is a finite element analysis tool that is running and Pro/Engineer.

Figure 9.10 shows the displacement and stress of the steel blade that is placed in the front of the plow. A load is applied to the edge of the blade. Figure 9.11 shows the load distribution on the chassis and Figure 9.12 shows the stress and displacement of the chassis due to these loads. Figure 9.13 shows the loads on the rear shaft and Figure 9.14 shows the stresses and displacement due to these loads. Figure 9.15 shows the loads on the wheel while Figure 9.16 shows the stresses and displacements due to these loads. The load results illustrated in Figures 9.10 through 9.15 show that the there would be no failure due to loading.

This type of analysis should be used as criteria when evaluating different designs. Designers should produce evaluation tables based on a part's performance under analysis. Designers could then change a part if the part has failed to fulfill its function.

An optimization for the concept based on the evaluation criteria, structural and functional analysis, and cost should be performed. In general, a mathematic model is developed to describe the different parameters influencing design. For example, a model to describe the cost based on the material choice, a model to describe the structural integrity, a model to describe the geometrical contribution to cost, and structural integrity, and so on. These parameters are then optimized to find the alternative that will provide the best performance with the lowest cost. The optimization techniques are mathematical algorithms that have an active area of research.

Many researchers are studying the building of a real-time design engine where the alternatives are modeled using Pro/E or other solid modelers. An analysis engine then would evaluate the alternative for its performance criteria if a change is needed based on the analysis. Then a parameter or group of parameters will be changed and the analysis run one more time. This procedure continues until an optimized solution is obtained.

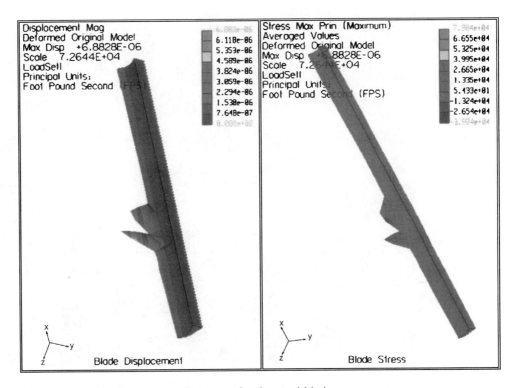

Figure 9.10 Displacement and stresses for the steel blade.

Figure 9.11 Loading on the chassis.

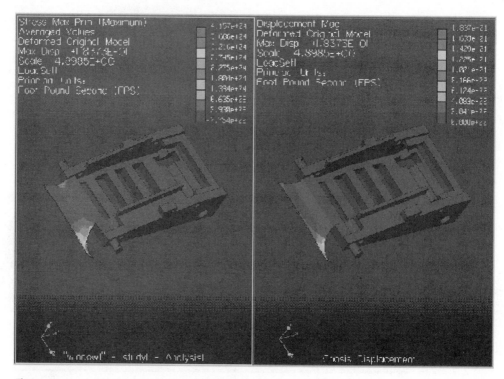

Figure 9.12   Displacement and stresses on chassis.

Figure 9.13   Loading on the rear shaft.

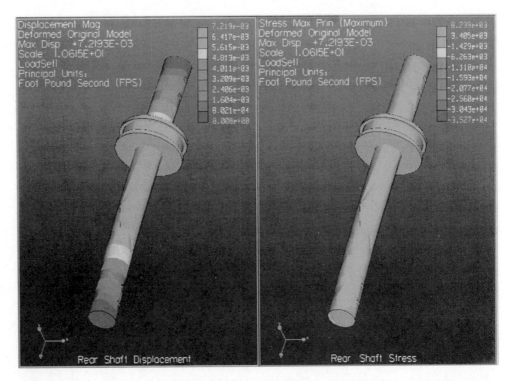

Figure 9.14 Displacement and stresses on rear shaft.

Figure 9.15 Loading on the large wheel.

Figure 9.16 Displacement and stresses on large wheel.

9.8 EMBODIMENT DESIGN: AUTOMATIC CAN CRUSHER

Details of the chosen design are sketched, and a bill of material is generated for that design.

9.9 PROBLEMS

9.9.1 Team Activities

1. Consider a regular stapler and a hole puncher.
   a. Draw a function tree.
   b. Sketch the individual elements that were used to perform the functions.
   c. Sketch the overall device (combine the elements to obtain the full device).
   d. Construct a limited bill of materials for the device.

2. Survey your school CAD software and list the capabilities if each software. Some of this software has a very small learning curve, which may allow you to produce a 3-D model of your product in a very short time. If you have engineering graphics experience from a previous course use that experience to check if the software has an analysis module that allows you to feed your model into the analysis module. Check the requirements of the analysis module.

3. In most architects' offices you will find a layout of future or current projects. What can you call these layouts?

4. List the bill of materials for a paper stapler.

9.10 SELECTED BIBLIOGRAPHY

1. Devon, R., and Engle, R. S. "The Effect of Solid Modeling Software on 3D Visualization Skill," *The Engineering Design Graphics Journal*, Spring, 1994.
2. Edwards, B. *Drawing on the Artist Within*. New York: Simon & Schuster Inc., 1986.
3. Haik, Y., and Kilani, M. *Essentials of Pro/E*. Pacific Gove, CA: Brooks/Cole, 2001.
4. Harrisberger, L. *Engineersmanship: The Doing of Engineering Design*. Belmont, CA: Wadsworth, Inc., 1982.
5. Juricic, D., and Barr, R. E. "The Place of Engineering Design Graphics in the Evolving Design Paradigm," Proceedings of the ASEE Annual Conference, Alberta, Canada, 1994.
6. Nee, J. G. "Freshman Engineering Design/Graphics Problem Status: A National Study," *The Engineering Design Graphics Journal*, Autumn, 1994.
7. Rodiguez, W. *Visualization*. New York: McGraw-Hill, 1990.
8. Roorda, J. "Visual Perception, Spatial Visualization and Engineering Drawing," *The Engineering Design Graphics Journal*, Spring, 1994.
9. Suh, N. P. *The Principles of Design*. Oxford University Press, 1990.
10. Ulman, D. G. *The Mechanical Design Process*. New York: McGraw-Hill, 1992.
11. Ulrich, K. T., and Eppinger, S. D. *Product Design and Development*. New York: McGraw-Hill, 1995.

10 Cost Analysis

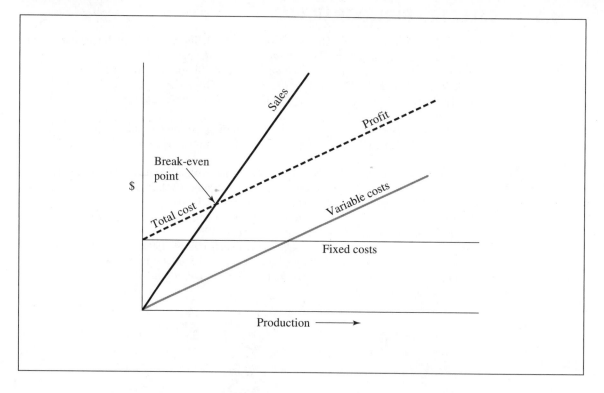

This break-even chart is designed to show, graphically, the profits and losses on the selling price and manufacturing costs of a product.

10.1 OBJECTIVES

By the end of this chapter, you should be able to

1. Discuss the different methods used in establishing cost analysis.
2. Use techniques introduced in this chapter to evaluate design cost.

One of the most essential tasks in engineering design involves estimating the production cost of a new product. Generally speaking, the lowest-cost design will be the most successful in a free marketplace. Understanding the elements that make up cost is essential. Furthermore, cost estimates are used

1. To provide a basis for estimating the selling price of the new product.
2. To distinguish between the different concepts based on material and manufacturing processes.
3. To provide information on the new product profit.
4. As a basis for redesigning certain elements to reduce cost.

It is important to generate a rough cost estimate as early as possible in the design process, and then to refine the cost estimate as soon as the embodiment phase is reached. As the design is refined, the cost estimate is also refined until the final product is produced.

Cost evaluation includes estimating the cost of building a plant, or the cost of installing a process within a plant to produce a product or a line of products. It also involves estimating the cost of manufacturing a part based on a particular sequence of manufacturing steps.

In design, it becomes important to find techniques that help reduce cost. Some of the cost reduction techniques that can be used are as follows:

1. Introduce new processes.
2. Take advantage of knowledge gained as experience increases; to allow for standardization of parts, materials, and methods.
3. Employ a steady production rate, when feasible.
4. Utilize all production capacity.
5. Have product-specific factories or production lines.
6. Improve methods and processes to eliminate rework and reduce the work in process and inventory.

It is important to avoid factors that tend to increase the costs of the design process. Factors that increase costs are as follows:

1. Incomplete product design specifications.
2. Redesign due to failures.
3. Supplier delinquency.

4. Management and/or personnel changes.
5. Relocation of facilities.
6. Unmet deadlines.
7. A product that has become too complex.
8. Technology and/or processes that are not developed as well as they were thought to be.
9. Inadequate customer involvement.
10. Disregarding needs.
11. Not subjecting a new product idea to competitive evaluation.
12. Not demonstrating that a new design can function properly under realistic conditions.
13. Designing without considering manufacturing processes.
14. Not continually improving and optimizing the product.
15. Relying on inspection (last minute test).

10.2   COSTS CLASSIFICATIONS

Each company or organization develops its own bookkeeping methods. In this section, the different classification divisions of costs are presented. Cost estimation within a particular industrial or governmental organization follows a highly specialized and standardized procedure, particular to the organization.

1. *Nonrecurring–recurring:* There are two broad categories of costs:
   a. Nonrecurring costs: These are one-time costs, which are usually called capital costs, such as plant building and manufacturing equipment.
   b. Recurring costs: These costs are direct functions of the manufacturing operations.
2. *Fixed–variable costs:* Fixed costs are independent of the rate of production of goods; variable costs change with the production rate.
   a. Fixed costs include
      i. Investment costs
         • Depreciation on capital investment
         • Interest on capital investment
         • Property taxes
         • Insurance
      ii. Overhead costs include
         • Technical service (engineering)
         • Nontechnical service (office personnel, security, etc.)
         • General supplies
         • Rental of equipment
      iii. Management expenses
         • Share of corporate executive staff
         • Legal staff
         • Share of corporate research and development staff

      iv. Selling expenses
- Sales force
- Delivery and warehouse costs
- Technical service staff

  b. Variable costs:
     i. Materials
     ii. Direct labor
     iii. Maintenance cost
     iv. Power and utilities
     v. Quality-control staff
     vi. Royalty payments
     vii. Packaging and storage cost
     viii. Scrap losses and spoilage

3. *Direct–Indirect:* A cost is called a direct cost when it can be directly assigned to a particular cost center, product line, or part. Indirect costs cannot be directly assigned to a product but must be spread over the entire factory.

  a. Direct costs include
     i. Material: Includes the expenses of all materials that are purchased for a product, including the expense of waste caused by scrap and spoilage.
     ii. Purchased parts: Components that are purchased from vendors and not fabricated inhouse.
     iii. Labor cost: Wages and benefits to the workforce needed to manufacture and assemble the product. This includes the employees' salaries as well as all fringe benefits, including medical insurance, retirement funds, and vacation times.
     iv. Tooling cost: All fixtures, molds, and other parts specifically manufactured or purchased for production of the product.

  b. Indirect costs include
     i. Overhead: Cost for administration, engineering, secretarial work, cleaning, utilities, leases of buildings, insurance, equipment, and other costs that occur day to day.
     ii. Selling expenses: marketing advertisements

10.3 COST ESTIMATE METHODS

The cost estimating procedure depends on the source of the components in the product. There are three possible options for obtaining the components:

1. Purchase finished components from a vendor.
2. Have a vendor produce components designed inhouse.
3. Manufacture components inhouse.

There are strong incentives to buy existing components from vendors. Cost is only one factor that determines whether a product or a component of a product will be made inhouse or bought from an outside supplier.

    If the quantity to be purchased is large enough, most vendors will work with the product design and modify existing components to meet the needs of the new product. If existing components or modified components are not available off the shelf, then they must be produced. In this case, a decision must be made regarding whether they should be produced by a vendor or made inhouse. This buy–make decision is based on the cost of the component involved, as well as the capitalization of equipment and the investment in manufacturing personnel.

The following table lists factors effecting the make–buy decision.

Reason to Make	Reason to Buy
Cheaper to make	Cheaper to buy
Company has experience making it	Production facilities are unavailable
Idle production capacity available	Avoid fluctuating or seasonal demand
Compatible and fits in production line	Inexperience with making process
Part is proprietary	Existence and availability of suppliers
Wish to avoid dependency on supplier	Maintain existing supplier
Part fragility requiring high packing	Higher reliability and quality
Transportation costs are high	

When manufacturing an element, the methods used to develop cost evaluation fall into three categories:

1. *Methods engineering (industrial engineering approach):* The separate elements of work are identified in detail and summed into total cost per part. Machined components are manufactured by removing portions of the material that are not wanted. Thus, the costs for machining are primarily dependent on the cost and shape of the stock material, the amount and shape of the material that needs to be removed, and how accurately it must be removed. An example for the production of a simple fitting from a steel forging is shown here:

Operations	Material	Labor	Overhead	Total
Steel forging	37.00			37.00
Set-up on milling machine		0.2	0.8	1.00
Mill edges		0.65	2.6	3.25
Set-up on drill press		0.35	1.56	1.91
Drill 8 holes		0.9	4.05	4.95
Clean and paint		0.3	0.9	1.2
Total	37.00	2.40	9.91	49.31

2. *Analogy:* Future costs of a project or design are based on past costs of a similar project or design, with due allowance for cost escalation and size difference. This method, therefore, requires a backlog of experience or published cost data.

3. *Statistical approach:* Techniques such as regression analysis are used to establish relations between system costs and initial parameters of the system: weight, speed, power, etc. For example, the cost of developing a turbofan aircraft might be given by

$$C = 0.13937 x_1^{0.7435} x_2^{0.0775}$$

where C is in millions of dollars, x_1 is the maximum engine thrust in pounds, and x_2 is the number of engines produced.

10.4 LABOR COSTS

The most accurate method of determining labor cost is to set up a staffing table to find the actual number of people required to run the process line. Be sure to account for standby personnel. Cost of operating labor includes fringe benefits. As an order of magnitude estimate of labor costs,

$$\frac{\text{Operating person-hour}}{\text{Tons of product}} = K\frac{\text{Numbers of process steps}}{(\text{tons/day})^{0.76}}$$

where

$K = 23$ for a batch operation

$K = 17$ for average labor requirements

$K = 10$ for an automated process

10.5 PRODUCT PRICING

Pricing can be defined as the technique used for arriving at a selling price for a product or service that will interest the customer while simultaneously returning the greatest profit. Different methods are available to assist in arriving at the optimum price for a product. One of these methods is known as the break-even chart.

10.5.1 Break-Even Chart

The break-even chart (Figure 10.1) is designed to show, graphically, the profits and losses on the selling price and manufacturing costs of the product. The manufacturing costs of a product include the following:

1. Material: Credit should be accounted for recycled scrap and by products.
2. Purchased parts
3. Labor
4. Tooling
5. Overhead

Material, purchased parts, and labor are variable costs that depend on the number of goods produced while tooling. Overhead is a fixed cost, regardless of the number of goods produced. The fact that the variable costs depend on the rate or volume of production while fixed costs do not leads to the idea of a break-even point. Determination of the production lot size needed to exceed the break-even point and produce a profit is an important consideration. To draw a break-even point chart, use the following steps:

1. Determine the variable costs per product. Variable costs depend on the number of products and include

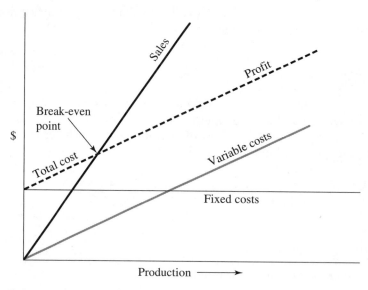

Figure 10.1   Break-even chart.

    a. Material
    b. Purchased parts
    c. Labor
2. Determine the fixed cost. Fixed cost does not depend on the volume or the number of products.
3. Draw, on the chart, the total cost as a function of the number of products. The total cost = fixed cost + variable cost.
4. Determine a sale price of the product. Draw the sales as a function of the number of products.
5. The intersection of the sales line with the total price is the break-even point.

10.5.2   Linear Programming

Linear programming is a tool that is used to represent a situation, in which an optimum goal is sought, such as to maximize profit and minimize cost. Mathematically, it involves finding a solution to a system of simultaneous linear equations and linear inequalities, which is optimized in linear form. An illustrative example is listed here for demonstration of this technique.

    A manufacturer of desk staplers and other stationary items produces two different stapler models (see Table 10.1). One is manual in operation, retails at \$12.50, and returns \$2.00 to profit. The other is automatic and sells for \$31.25, returning \$5.00 to profit. The manufacturer is concerned with how to utilize his production facilities in order to maximize profits. One constraint placed on the maximizing of profits can be written as follows:

$$P = 2Q_M + 5Q_A$$

TABLE 10.1   Stapler Models

	Manual	**Automatic**
Selling price	31.25	12.5
Profit	$2	$5
Constraints		power supply 200 per day
Labor	18 person-minutes	54 person-minutes
Adjustment	3 person-minutes	5.4 person-minutes
Inspection	1 person-minute	1.5 person-minutes

where P is profit function, Q_M is the quantity of manual staplers, and Q_A is the quantity of automatic staplers.

Assume that the manufacturing operation is such that labor skills and plant facilities are completely interchangeable; thus, varying quantities of manual and automatic staplers may be produced on the same day. Also, assume that the present demand is such that the company can sell as many as it can produce. There are certain production restraints. Most of the major components are prepared on separate assembly lines, prior to moving to the final assembly area. There is a sufficient inventory of components for both staplers to satisfy daily production, with the exception of solenoids for the automatic model, which has a daily production capacity of 200. This constraint may be written as

$$Q_A \leq 200$$

Among other duties of employees, fabrication and assembly requires 18 person-minutes for each manual model and 54 person-minutes for each automatic stapler. The plant employs 45 people for this task, on an 8-hour shift. Since there is only one shift, the capacity is $8 \times 45 \times 60$, or 21,600 person-minutes per day. This may be defined mathematically by

$$18Q_M + 54Q_A \leq 21,600$$

The final adjustment in the manual stapler requires 3 person-minutes on the assembly line, whereas the automatic model requires 5.4 person-minutes. This phase requires 6 people, and the daily capacity is $8 \times 6 \times 60$ or 2880 person-minutes. This constraint is expressed as

$$3Q_M + 5.4Q_A \leq 2880$$

Final inspection for workability requires 1 person-minute for manual models and 1.5 person-minutes for the automatic model. There are 4 half-time inspectors assigned to this task; the inspection capacity is $8 \times 2 \times 60$ or 960 person-minutes. The inspection constraint is

$$Q_M + 1.5Q_A \leq 960$$

The constraints equations are as follows:

$$P = 2Q_M + 5Q_A$$

$$Q_A \leq 200$$

$$18Q_M + 54Q_A \leq 21{,}600$$

$$3Q_M + 5.4Q_A \leq 2880$$

$$Q_M + 1.5Q_A \leq 960$$

Plotting these equations yields Figures 10.2 and 10.3.

Linear programming is used to find the isoprofit line (profit is the same along the line) that is farthest away yet still contains at least one point that is technically feasible.

A combination of break-even and linear programming approaches can be used to set a price for a product. The linear programming approach can be used to determine the optimum number of goods; then that number may be used on the break-even chart to determine the selling price. The actual setting of a price resides at the heart of business practice, but describing it briefly is not easy. A key issue in establishing a price involves the reaction of the competition and the customer. When competition is fierce, the sales force may push hard for price reductions, claiming that they can make up the lost revenues by increasing the volume. Special offers, such as promotional discounts, premiums, extras, trade-ins, volume discounts, and trade-in allowances, are often used.

Figure 10.2 Constraints equation without profit.

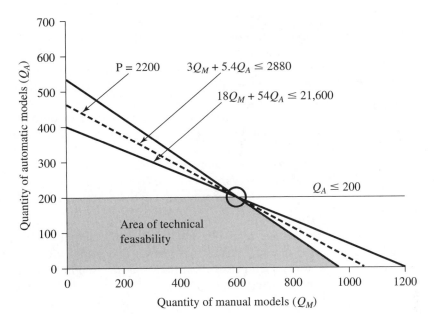

Figure 10.3 Constraints equation with profit.

10.6 PROBLEMS

10.6.1 Team Activities

The objective of this exercise is to use computer programming to write a program that can find the cost of different alternatives for your design project. Manufacturing cost is an important factor in evaluating the different alternatives. Some alternatives will be eliminated because of their high manufacturing cost.

Outline
In engineering design projects, students are expected to build a device by the end of a limited time period. With a limited amount of money, students need to figure out the manufacturing costs of different designs before they actually start building these devices. A manufacturing cost is a key determinant of the economic success of a product. When computing the manufacturing cost, one can divide the cost into two broad categories: (1) fixed costs and (2) variable costs. Fixed costs are incurred regardless of the number of products that are manufactured, whereas variable costs depend on the amount of goods that are produced. In student design projects, students are not expected to buy major tools or rent warehouses for production; hence the fixed costs and variable costs in this case may use the following components:

1. Fixed Costs
   a. *Tooling cost:* Tooling costs can include buying or renting. Renting depends on the charge per unit time, regardless of the number of units produced. Buying in this case will be for this project only, and it will be assumed that it is paid for in full (i.e., no interest and no depreciation).
   b. *Setup cost:* This is the cost required to prepare equipment for a production, and it is fixed regardless of the number of units produced.
2. Variable Costs
   a. *Material cost:* See Figure 10.4 (on the next page).
   b. *Processing cost:* Varies with the type of manufacturing equipment used, and includes charges for both machine time and labor.

Programming Requirements
The program should be readable; all items from the terminal are sorted into "ready" and "need-manufacturing" categories. The information should be fed by the user. Ready items need prices from the vendor, while items that need manufacturing require the following calculation:

1. User inputs the following information:
   a. Mass of the material (see provided table)
   b. Total number of units of this type
   c. Setup cost
   d. Tooling cost
   e. Processing cost
2. The computer calculates the manufacturing cost for each element.

The output should be in the format shown in Table 10.2.

TABLE 10.2 Concept Number 1, Attach a Figure. Computer Output of Cost Calculation for Each Alternative

Component	Material	Ready	Part Number	Vendor	Fixed Cost	Variable Cost	Total

10.6.2 Individual Activities

1. Name three different cost classifications, and state the differences among them.
2. What are the three cost estimate methods, and what are the differences among them? Which method is more appropriate for your design project and why?
3. What is the break-even chart?
4. Using the break-even method, find break-even points if
   a. The fixed cost is $12,000.
   b. The variable cost is $17.95 per unit.
   c. The sale price is $29.50 per unit.

Figure 10.4 Material cost. Range of costs for common engineering materials. Price ranges shown correspond to various grades and forms of each material, purchased in bulk quantities (1994 prices). (Adapted from David G. Ullman, *The Mechanical Design Process*, McGraw-Hill, New York, 1992.)

5. What would be the break-even in the previous question if the fixed cost is the same but the variable cost becomes $29.50 and the selling price becomes $37.50?

6. The following data are for a company that produces washers and Dryers.
   - Dryers retail for $198 and contribute $15 to profit.
   - Washer retail for $499.95 and contribute $45 for profit.
   - The washer blade is limited in production capacity to 50 blades, while all other components have no limits.
   - Chassis assembly requires 6 person-hours for each dryer set and 18 person-hours for each washer. The plant employs 225 workers for an 8-hour shift to perform chassis assembly operations.
   - A dryer requires 1 person-hour on the assembly line, a washer set 1.6 person-hours. There are 30 people on a single 8-hour shift assigned to assembly.
   - Final inspection requires 0.5 person-hour for a dryer and 2.0 person-hours for washer. The plant employs 20 full-time inspectors and one part-time employee for 2 hours per day.
   a. Write a linear programming model.
   b. What is the optimum number of dryers and washers?
   c. Calculate the maximum profit.
   d. The linear programming and break-even approaches are used to find the selling price for both the dryer and washer, based on $10,000 fixed cost for both dryer and washer with a variable cost of $160 for the dryer and $330 for the washer. Determine how many days will be required until the company starts making a profit for the washers and the dryers. Use $198 and $500 as selling prices for the dryer and the washer, respectively.

10.7 SELECTED BIBLIOGRAPHY

1. Dhillon, B. S. *Engineering Design: A Modern Approach.* Toronto: Irwin, 1995.
2. Dieter, G. *Engineering Design.* New York: McGraw-Hill, 1983.
3. Dym, C. L. *Engineering Design: A Synthesis of Views.* Cambridge, UK: Cambridge University Press, 1994.
4. Hill, P. H. *The Science of Engineering Design.* New York: McGraw-Hill, 1983.
5. Pugh, S. *Total Design.* Reading, MA: Addison-Wesley, 1990.
6. Ray, M. S. *Elements of Engineering Design.* Englewood Cliffs, NJ: Prentice Hall, 1985.
7. Radcliffe, D. F., and Lee, T. Y. "Design Methods used by Undergraduate Students," *Design Studies*, Vol. 10, No. 4, pp. 199–207, 1989.
8. Suh, N. P. *The Principles of Design.* New York: Oxford University Press, 1990.
9. Ulman, D. G. *The Mechanical Design Process.* New York: McGraw-Hill, 1992.
10. Ulrich, K. T., and Eppinger, S. D. *Product Design and Development.* New York: McGraw-Hill, 1995.
11. Vidosic, J. P. *Elements of Engineering Design.* New York: The Ronald Press Co., 1969.
12. Walton, J. *Engineering Design: From Art to Practice.* New York: West Publishing Company, 1991.

Part II
Sample Design Projects

11 Sample Design Projects

In this chapter, we will discuss some sample design projects, including one for this solar toy car.

11.1 DESIGN PROJECT RULES

This chapter introduces sample design projects. The following rules apply to all design projects. Specific rules will also be listed for each project.

1. Students will apply a systematic design process to build the prototype.
2. Students will develop an objective tree to clarify the need statement and prioritize the objectives laid out in the problem statement. Students may add features that may not be listed clearly in the problem statement but will give an additional advantage to the proposed design. The objective tree is a mechanism to organize the objectives or design criteria in a tree fashion and combine different objectives under a higher objective.
3. Students will conduct a market analysis. In the market analysis students will conduct surveys to test and collect the customers wishes. Students will also conduct a literature search of all industry using the Web and evaluate the future market. A patent search should also be conducted to prevent preoperative patent violations. The large market players should be identified. Study of the industry trends in the last 5 years should be conducted.
4. Students should develop a specification table. The specification table should recognize which of the specifications is a demand and which is a wish. The demands are laid out in the problem statement, while the wishes are obtained from the objective tree. A house of quality can be developed to enable students to better understand the objectives they have set up.
5. Students should develop a functional analysis of the project. In the function analysis students will lay down the functions that involve in a working model.
6. Students will develop a morphological chart that includes all functions and mechanisms able to perform that function.
7. Students will generate several design alternatives by combining the mechanisms in the morphological chart in accordance with the function layout.
8. Students will evaluate the different alternatives by using the Pugh method to identify the best design.
9. Students will generate a detailed drawing with assembly of their design. Solid modeling will be utilized before a prototype is generated. Students will optimize their design and modify their design on Pro/E solid drawings.
10. Students will build a mock-up of the design to check the conceivability and the manufacturability of the different components.
11. Students will generate a bill of material and identify the vendors, cost, and different part alternatives.
12. Students will calculate the cost and use the break-even point to find the selling price of their product.

13. Students will use the machine shop to build their design project.

14. Students will be trained in the different design tools. Students will use the Gantt, CPM, or PERT methods for scheduling the different events in the design process.

15. Students will be introduced to the different manufacturing processes during the course of the design.

16. The device must be a stand-alone unit. No assistance from the operator will be allowed after actuation.

17. Any device that is available through a retailer will be disqualified. The device must be designed and constructed from the component level by team members.

18. Groups will be given a 2-minute warning before they compete. They must report to the competition site within this time or they will be disqualified.

19. Judges and/or instructors will disqualify any device that appears to be a safety hazard.

20. The total cost of the device should not exceed the given budget ($200). A tolerance of 10% may be allowed after discussion with the instructor.

21. A complete breakdown of all costs must be included in the report. Receipts must be supplied

22. Teams are to comprise four members, to be selected and decided on by the students.

23. The Rhomberg method will be used by students to evaluate each team members contribution. Student evaluations of team members will be kept confidential and are to be turned in weekly. These evaluations will be considered for grading.

24. The first phase includes the objective tree, function tree, market analysis, ideas, concepts sketches, schedules, and material selection.

25. The second phase includes evaluation, optimization, analysis, modeling, prototype/mock-up, and marketing.

26. Teams may choose to exceed the budget by 10% (the overall cost should never exceed $220).

27. The technical report and marketing brochures must be turned in.

28. All devices will be collected and impounded two days before the competition (i.e., students will not be allowed to take the devices away until the completion of all presentations).

29. The top three projects will be showcased by the department. Commendations and prizes will be awarded. Judging will be based on the following criteria:
    a. Performance
    b. Appearance
    c. Safety
    d. Other general engineering principles

30. The judges'/instructor's decisions at all times will be final. Appeals by students will be heard, but once a decision is reached it will be considered as final and no further appeals will be tolerated.

31. Students are reminded of the general rules and honor codes of their respective institutions.

32. Students are encouraged to be creative and not restrict themselves to ideas that are currently available. This exercise should be a positive learning experience for the students; competitiveness is encouraged. However, the spirit of fairness must always prevail, and students are requested to abide by the rules and decisions of the judges and instructor.

33. The instructor reserves the right to change/add rules to fit student learning needs. Students will be informed of the changes when they occur.

11.2   ALUMINUM CAN CRUSHER

11.2.1   Objective

Design and build a device/machine that will crush aluminum cans. The device must be fully automatic (i.e., all the operator needs to do is load cans into the device). The device should switch on automatically, crush the can automatically, eject the crushed can automatically, and switch off automatically (unless more cans are loaded).

11.2.2   Specifications

- The device must have a continuous can feed mechanism.
- Cans should be in good condition when supplied to the device (i.e., not dented, pressed, or slightly twisted).
- The can must be crushed to one-fifth of its original volume.
- The maximum dimensions of the device is not to exceed $1.5 \times 1.5 \times 1$ feet.
- Performance will be based on the number of cans crushed in one minute.
- Elementary school children (kindergarten up) must be able to operate the device safely.
- The device must be a stand-alone unit. No assistance from the operator will be allowed after actuation.
- Any device that is available through a retailer will be disqualified. The device must be designed and constructed from the component level by team members.

11.3   COIN SORTING CONTEST

11.3.1   Objective

Design and build a machine/device that will sort and separate different types of coins. A bag of assorted coins will be emptied into the device and the contestants will be given one minute to sort and separate as many coins as possible. Coins will be disbursed in a bag containing an assortment of U.S. coins (pennies, nickels, dimes, and quarters) and other foreign coins. The device must be able to sort out and separate only the U.S. coins. The volume of coins that the bag will contain is approximately one-eighth of a cubic foot. The coins will be deposited into the device at the start time. The number of coins sorted and separated will be counted at the end of time (60 seconds). Each incorrect coin separated will be penalized. Each coin that is physically damaged by the device will be penalized. Coins that remain in the internal body of the device at the end of time will not be counted. Each round will be run independently (i.e., the only opponent is the clock). Each team will compete twice on two different cycles, based on random selection. The better of the two performances will be considered in the contest.

11.3.2   Constraints

The total cost of the device should not exceed the given budget ($200). A tolerance of 10% may be allowed after discussions with the instructor. A complete break-down of all costs must be included in the report. Receipts must be supplied. The device must be a stand-alone unit. No assistance from the operator will be allowed after actuation. The operator will be allowed to power the device either manually or by any other means. Any device that is available through a retailer will be disqualified. The device must be designed and constructed from the component level by team members. If there are any questions, the instructor should be consulted for clarification.

11.4   MODEL (TOY) SOLAR CAR

11.4.1   Objective

Design and build a model solar car that can be used as a toy by children. The primary source of energy is solar.

The following are some methods to stimulate your thinking; you are not limited to these methods.

- Direct solar drive.
- Battery storage (Students should note that on the day of the competition, the battery must be completely discharged and they will be given half an hour to charge the battery, using solar energy.)
- Air or steam engine (Students will be allowed to use photocells, lenses, etc. for the purpose of heating the air, or steam generation.)

11.4.2   Specifications

- The car will be required to travel 100 feet in a straight line. Extra credit will be given to cars that are remote controlled. The final competition will take place in a school parking lot.
- Performance will be based on the car that travels the distance of 100 feet in the least amount of time.
- The car must travel a distance of at least 10 feet or it will be considered damaged. Damaged cars will get a failing grade in the performance section of the competition.

- The dimensions of the car are limited to the following maximum values:
  Length: 1 foot

  Width: 6 inches

  Height: 6 inches
- The car must be safe for children to use.
- Any car that is available through a retailer will be disqualified. The car must be designed and constructed from the component level by team members.

11.5 WORKSHOP TRAINING KIT

The Stirling engine kit that is used for a Mechanical Engineering Tools class is becoming more expensive and difficult to obtain on time. The goal of using that kit is to provide students with hands-on experience using the machine shop and at the same time produce a machine that is able to convert heat into work. This project calls for students to design another kit that can be designed and manufactured at the College of Engineering and serve the same goal of providing hands-on experience and building a machine that is able to convert one form of energy to another. Design and build a machine shop training kit that can be used for a Mechanical Engineering Tools class.

11.5.1 Specifications

- The kit must have teaching and training value to the students. The kit must be innovative and utilize an engineering principle to produce work (such as the first law of thermodynamics).
- The kit must have identified training values for machine shop tools.
- The dimensions of the kit when assembled are limited to the following maximum values: Length: 1 foot; width: 6 inches; and height: 6 inches.
- The kit, once assembled, must be safe to use for students in kindergarten through grade 12.
- Any kit that is available through a retailer will be disqualified. The kit must be designed and constructed from the component level by team members.

11.6 SHOPPING CARTS

As technology advances and new materials are synthesized, certain products are not modified and modernized. Among those are the shopping carts that being used in grocery stores. As many of you may have observed, there is a tendency to conserve parking space by not designating a return cart area. Leaving carts in the parking lots may lead to serious accidents and car damage. Furthermore, many customers do not fill their carts when shopping; however, they do not like to carry baskets. Other customers like to sort products as they shop.

Design and build a new shopping cart that can be used primarily in grocery stores. The shopping cart should solve the common problems in the available carts.

11.6.1 Specifications

You must adhere to the following rules:

- Conduct surveys to measure customer needs for shopping carts.
- The shopping cart should be safe for human operation.
- The dimensions of the cart should match existing carts.
- The cart should be easy to operate by children of age 7 and older and senior citizens.
- The cart should have features that accommodate children while shopping.
- The cart should be able to accommodate large and small items.
- Any cart that is available through a retailer will be disqualified.

11.7 MECHANICAL VENTS

Most houses have vents that open and close manually without any central control. Cities across the country advise the use of such vents to save energy. In most cases, household occupants do not use the entire house at the same time; the tendency is to use certain rooms for a long time. For example, the family room and dining room may be used heavily, while the living room and kitchen are used at certain hours of the day. To cool or heat a room, the system needs to work to cool or heat the entire house. Energy saving can be enhanced if the vents of unused rooms are closed; this will push the hot/cold air to where it is needed most and reduce the load of the air conditioning system.

Design and build a remotely operated ventilation system that will be connected to an existing ducting system. The house layout will be given upon request.

11.7.1 Specifications

You must adhere to the following rules:

- The system must be safe and easily adapted to existing ventilation systems.
- The system must work for at least seven exits.
- The venting system must accommodate high and low operating temperatures.
- The system must be easy to operate by household members.
- Central units of operation are encouraged as well as remote controlled units.
- Any system that is available through a retailer will be disqualified. The system must be designed and constructed from the component level by team members.

11.8 ALL TERRAIN VEHICLE

The objective of this project is to design and build a model for an all terrain vehicle (ATV). The size of the model should not exceed $(6 \times 6 \times 6)$ in^3. The power of the model vehicle is flexible, but safety must prevail. Success of this model design may lead to a new market for ATVs, such as

- A toy for children.
- A demonstration model for object collection on other planets.
- A demonstration model for law enforcement and military purposes.

The designed model must be able to handle slopes of 45 to 60 degrees, go over rocks that are the height of the wheels, and move in mud and dry dirt.

11.9 POCKET-SIZED UMBRELLA

As technology advances, the public is searching for more convenient products to replace existing products. In this project you are requested to design and build an umbrella that can fit in a normal pocket when folded. The following must be adhered to when designing the new umbrella:

- The umbrella must be able to cover the prospective customer from heavy rain.
- The response time to open and fold the umbrella must be reasonable and close to the response time of an existing full-size umbrella.
- The weight of the umbrella should be reasonable to fit in a pocket with damaging the pocket.

11.10 MODEL OF THERAPEUTIC WHEELCHAIR

In homes of the elderly and infirm and in and medical therapy clinics, a nurse is requested to help residents/patients walk on a daily basis. Most of the residents and patients are unable to walk alone. They use standard wheelchairs most of the day. Because of cost factors, only one nurse can be assigned to each patient. The patient usually leaves the wheelchair and walks away from it with the help of the nurse. In this situation a problem may arise if the patient needs the wheelchair urgently while he and the accompanying nurse are away from it. The nurse cannot leave the patient unattended to bring the chair; neither can she carry the patient back to the wheelchair. You are required to design and build a prototype wheelchair that will provide the necessary solution for the nurse. The model should not exceed $(1 \times 1 \times 1)$ ft^3. You need to consider that the patient/resident may be walking outdoors or indoors. Usually the nurse walks the patient within a 100-ft diameter. In urgent situations the nurse will need the chair to be available within one minute.

11.11 DISPOSABLE BLOOD PUMP

In blood treatment applications there is a need for a small disposable pump to reduce the interaction of the patient blood with operators and equipment. The pump can be placed along with other disposable fixtures. You are required to design and build a pump with the flowing specifications:

- All blood-contacting parts are disposable.
- The pump can be mass produced to reduce cost.
- The pump must be able to achieve a variable flow rate from 0 to 100 m/min with a volumetric accuracy of $\pm 5\%$.
- The pressure head is between -100 mm Hg to 100 mm Hg.
- Disposable size is not to exceed $1 \times 1 \times 10$ in^3. The overall fixture must not to exceed $(1 \times 1 \times 1)$ft^3.

11.12 NEWSPAPER VENDING MACHINE

The specific problem with existing newspaper dispensers is the ease of stealing the papers in the machine when the door is opened. Theft from existing newspaper vending machines is a source of revenue loss as well as frustration for newspaper companies today. Another limitation of the current dispensers is that they can only handle one newspaper for each dispenser, which makes them inefficient spacewise. You are required to design and build a newspaper dispenser with the following specifications:

- Can handle at least three different newspapers.
- Dispenses one newspaper at a time based on customer selection.
- Newspaper companies make most of their profit from advertisements and not from newspaper sales, so they will be unlikely to purchase machines that cost more than the current ones.
- Familiarity of the design to customers is crucial, since customers (newspaper companies) will be unlikely to purchase newspapers from designs that deviate from existing designs.
- The design size should conform to current newspaper dispenser standards, especially with regard to height.

11.13 PEACE CORPS GROUP PROJECTS

The class is assigned to a Peace Corps mission to village of an average population of 250 somewhere in the universe where electricity is not available. The village needs the teams' help to design mechanical instruments that will help the village to function.

11.13.1 Projects

The villagers require help from the team for the following:

- A mechanism to pump water. Their current technology is an open well with a bucket. Because of health reasons, it is best to cover the well and design a mechanical pump. The power for the pump could be provided manually or animal powered. Pumps are needed for both the drinking and irrigation system. Because of limitations in material resources, the pump must be used for both irrigation and for drinking water. (Assume the water is drinkable.) The flow rate needs to be controllable. Drinking water demand is much less than that of the irrigation system (about 1/200). The well is about 100 m deep (Don't ask me how they reached that depth.)
- A mechanism to grind wheat to produce flour. The current technology is that each household has their own manual grinder composed of two rough, heavy disks. The disks are aligned around a central hole where the wheat is fed. This current technology is time consuming and a major waste of resources. It is best to set up a central location for this task that every household could use. The new mechanism must save time and could be a one-man operation.

- A mechanism to help in seeding agricultural land. The current technology utilizes a shuffle and digger. If you have ever planted seeds by hand, you will know that this is a labor-intensive job. Average land per household is about 300 acres (30 households). The system needed should be energy independent. Please do not humiliate the people by suggesting a simple attachment to a horse.
- It is assumed that with the technology developed to pump the water and seed the ground, the yield will become 10 times higher than before. The task now is to develop a mechanism to collect the wheat and vegetables. Two systems are needed:
  (a) One for wheat collection and packaging; and
  (b) Another for vegetable collection and packaging.
     The average temperature is about 80°F during a year.
- A system to generate electrical energy would be of interest to the village. It is very clear one source will not be sufficient. They have a lot of wood at their disposal. Wind is more viable than solar because of material.
- A time and calendar system viewable by the whole village. Currently time is known by the shadow length of a well-designed wood column. What is needed is a mechanism that tells the time, day, month and year.

11.13.2   Materials

The village is out of touch when it comes to the availability of material. You are allowed to have $100 worth of material shipped to you. Be wise in using the material. Remember, lumber is available by the tons.

11.13.3   Machining

Machining would be the greatest challenge. There is no machine shop available. What is available is manual driven wood shaping equipment.

11.13.4   Deliverables

1. Teams will show three different alternatives of their designs using Pro/E or another CAD tool: Teams will apply a systematic design process in reaching three alternative designs using CAD tools. Teams are encouraged to be creative and not restrict themselves to ideas that are currently available.
2. Teams will show the analysis of the three alternatives as part of the evaluation process. In addition to the design objectives, the analysis is an important part in reaching a design decision. Details of the analysis are required.

Part III
Design Laboratories

The labs presented here are intended as tools to enhance students' participation and to reinforce concepts covered in the book. The author understands that some of these labs may not fit instructors' classes and there could be a need for additional or different labs. Thus the author is developing a Web page that invites instructors and design practitioners to submit material that can be shared with others. The Web page will maintain a database of all approved submissions. The Web page will be hosted by the author at http://www.eng.fsu.edu/~haik/design.

Lab 1: Ice Breaking—Forming Teams

In this chapter, we will discuss such team-building skills as brainstorming. Each person thinks and behaves in preferred ways that are unique to that individual. Harnessing each individual's thinking style can lead to productive brainstorming sessions.

12.1 PURPOSE

Thought processes have been studied since ancient history. Several models have been proposed on how the human brain works. One of the well-known models is the Herrmann model, which divides the brain into a four quadrant brain dominance model. These quadrants are located in the left and right hemispheres (left and right brain). The left cerebral hemisphere [A quadrant] in Herrmann's model is associated with logical, analytical, and quantitative thinking. The left limbic system [B quadrant] is associated with sequential, organized, and detailed thinking. The right cerebral hemisphere [D quadrant] is associated with visual, intuitive, and innovative thinking. The right limbic system [C quadrant] is associated with emotional, sensory, and interpersonal thinking. This lab introduces these models and provides a tool with which students can better understand their thought processes. When students understand their own thought processes and those of others, they work better in a group.

The objectives of this lab are as follows:

1. Introduce the model of the brain developed by Ned Herrmann.
2. Present a preliminary self-test to quantify the strength of the four quadrants.
3. Introduce team formation. After the students test which thinking preference they may have, it becomes a tool when the teams form to make a full brain (i.e., one student from A, another from B, and so on).

12.2 THEORY

Each person thinks and behaves in preferred ways that are unique to that individual. These dominant thinking styles are the results of the native personality interacting with family, education, work, and social environments. People's approaches to problem solving, creativity, and communicating with others are characterized by their thinking preferences. For example, one person may carefully analyze a situation before making a rational, logical decision based on the available data. Another may see the same situation in a broader context and look for several alternatives. One person will use a very detailed, cautious, step-by-step procedure. Another has a need to talk the problem over with people and will solve the problem intuitively. Ned Herrmann's metaphorical model divides the brain into left and right halves and into cerebral and limbic hemispheres, resulting in four distinct quadrants. The Hermann profile results are value neutral. There are no right or wrong answers.

The questionnaire presented in the next section will help you investigate your own thinking style. Questionnaires like this one are available from specialized companies such Ned Herrmann's for a nominal charge per student. The idea here is to have this activity and questionnaire as a preliminary step for people who have more interest in finding more and sufficient for people who just want some idea. The questions in this questionnaire are based on data available in Lumsdaines' book[1] and material distributed through the NSF (National Science Foundation) workshop on introductory engineering design at the Central Michigan University, 1999, presented by Frank Maraviglia.[2]

12.3   PROCEDURE

In the space to the right of each statement, write the numbers 4, 3, 2, or 1, where a 4 indicates most likely and 1 indicates least likely. Use each number only once for each question. There is no right or wrong answer; answer to the best of your knowledge and experience.

1. In a project setup you prefer to perform the following function:
   a. Looking for data and information   [   ]
   b. Developing a systematic solution and directions   [   ]
   c. Listening to others and sharing ideas and intuitions   [   ]
   d. Looking for the big picture and context, not the details   [   ]

2. While solving your homework you
   a. Organize the information logically in a framework but not down to the least detail   [   ]
   b. Do detailed homework problems neatly and conscientiously   [   ]
   c. Motivate yourself by asking why and by looking at personal meaning   [   ]
   d. Take the initiative in getting actively involved to make learning more interesting   [   ]

3. In studying a new material you prefer
   a. Listening to the informational lecture   [   ]
   b. Testing theories and procedures to find flaws and shortcomings   [   ]
   c. Reading the preface of the book to get clues on the purpose   [   ]
   d. Doing simulations and ask what-if questions   [   ]

4. Do you like
   a. Reading textbooks   [   ]
   b. Doing lab work step by step   [   ]

[1]Lumsdaine, E., and Lumsdaine, M. *Creative Problem Solving*. New York: McGraw-Hill, 1995.
[2]NSF Sponsored Workshop on Introductory Engineering Design, J. Nee (editor). Central Michigan University, 1999.

   c. Talking about, seeing, testing, and listening [ ]
   d. Using visual aids instead of words [ ]

5. In studying new material you like to
   a. Analyze example problems and solutions [ ]
   b. Write a sequential report on the results of a lab experiment [ ]
   c. Do hands-on learning by touching and seeing [ ]
   d. Take an open-ended approach and find several solutions [ ]

6. When and after solving a problem you
   a. Think through ideas rationally [ ]
   b. Find practical uses of knowledge learned; theory is not enough [ ]
   c. Use group study opportunities and group discussion [ ]
   d. Appreciate the beauty in the problem and the elegance of the solution [ ]

7. Do you prefer to
   a. Do research using the scientific method [ ]
   b. Use computers with tutorial software [ ]
   c. Use group study opportunities and group discussion [ ]
   d. Lead brainstorming sessions in which wild ideas are accepted [ ]

8. In project execution, which of the following you prefer to do?
   a. Make up a hypothesis and then test it to find out if it's true [ ]
   b. Plan projects, schedule and execute them according to a set time [ ]
   c. Keep a journal to record what you have seen in the experiment [ ]
   d. Experiment and play with ideas and possibilities [ ]

9. Generally speaking, you absorb new ideas best by
   a. Applying them to concrete situations [ ]
   b. Concentrating and conducting careful analysis [ ]
   c. Contrasting them with other ideas [ ]
   d. Relating them to current or future activities [ ]

10. When you read self-help articles you pay most attention to
   a. The ideas that are drawn from the information [ ]
   b. The truth of the finding backed up by information [ ]

    c. Whether or not the recommendations can
      be accomplished   [   ]

    d. The relation of the conclusion to your
      experience   [   ]

11. When you hear people arguing you favor the side that
    a. Presents ideas based on facts and logic   [   ]
    b. Expresses the argument most forcefully and
      concisely   [   ]
    c. Reflects your personal opinion   [   ]
    d. Projects the future and shows the total picture   [   ]

12. When you make a new choice you rely most on
    a. Reality and the present rather than future
      possibilities   [   ]
    b. Detailed and comprehensive studies   [   ]
    c. Talking to people   [   ]
    d. Intuition   [   ]

13. When you approach a problem you are likely to
    a. Try to relate to a broader problem or theory   [   ]
    b. Try to find the best procedure for solving it   [   ]
    c. Try to imagine how others might solve it   [   ]
    d. Look for ways to solve the problem quickly   [   ]

Add up the columns vertically and register   I   II   III   IV
the numbers in the space provided: [   ][   ][   ] [   ]

In items 14 to 17 you will find four lists. Check the number of items in each list that you enjoy doing. Add up the number of checked items and write the number in the space provided for each item.

14. Total: [   ]
    — Looking for data and information; doing library searches
    — Organizing information logically in a framework
    — Listening to informational lectures
    — Reading textbooks
    — Studying sample problems
    — Thinking through ideas
    — Making up a hypothesis and then testing it to find out if it is true
    — Judging ideas based on facts, criteria, and logical reasoning
    — Doing technical and financial case studies
    — Knowing how much things cost
    — Using computers for math and information processing

15. Total: [   ]
    — Following directions carefully, instead of improvising
    — Testing theories and procedures to find flaws and shortcomings
    — Doing lab work step by step
    — Listening to detailed lectures

— Taking detailed comprehensive notes
— Studying according to a fixed schedule in an orderly environment
— Making up a detailed budget
— Practicing new skills through frequent repetition
— Writing a how-to manual about a project
— Using computers with tutorial software
— Finding practical uses of knowledge learned

16. Total: [   ]
— Listening to others and sharing ideas
— Keeping a journal to record feelings and spiritual values
— Studying with background music
— Respecting others' rights and views; people are important, not things
— Using visual clues to make use of body language
— Doing hands-on learning by touching and using a tool or object
— Watching drama movies over adventure movies
— Learning through sensory input (i.e., moving, feeling, smelling, testing, listening)
— Motivating yourself by asking why and by looking for personal meaning
— Traveling to meet other people and learn about their culture

17. Total: [   ]
— Looking for the big picture and context, not details, of a new topic
— Taking the initiative in getting actively involved to make learning more interesting
— Doing open-ended problems and finding several possible solutions
— Thinking about trends
— Thinking about the future and making up long-range goals
— Admiring the elegance of inventions, not the details
— Synthesizing ideas and information to come up with something new
— Relying on intuition to find solutions
— Predicting what future technology may look like
— Looking for alternative ways of accomplishing a solution
— Using pictures instead of words when learning

To find out the totals you need to

1. Multiply the total number in 14 by 4 and then add the results to the I bracket. Once you have the total sum, call it A.
2. Multiply the total number in 15 by 4 and then add the results to the II bracket. Once you have the total sum, call it B.
3. Multiply the total number in 16 by 4 and then add the results to the III bracket. Once you have the total sum, call it C.
4. Multiply the total number in 17 by 4 and then add the results to the IV bracket. Once you have the total sum, call it D.

12.4 SAMPLE RESULT

The following is a sample of one student's questionnaire results:

[1] [3] [4] [2]
[3] [4] [1] [2]
[4] [3] [1] [2]
[2] [1] [4] [3]
[4] [1] [3] [2]
[4] [2] [3] [1]
[4] [2] [3] [1]
[3] [4] [1] [2]
[3] [4] [2] [1]
[4] [1] [3] [2]
[3] [2] [4] [1]
[4] [1] [3] [2]
[3] [4] [1] [2]

I   II   III IV

- Total [42] [32] [33] [23]
- 14 Total [7]
- 15 Total [5]
- 16 Total [4]
- 17 Total [4]

The sum is expressed as

$$A = 4 * 7 + 42 = 70$$

$$B = 4 * 5 + 32 = 52$$

$$C = 4 * 4 + 33 = 49$$

$$D = 4 * 4 + 23 = 39$$

To express the finding graphically, use Excel spreadsheet. Then use the radar chart within Excel to express the values graphically, as shown in Figure 12.1. People with an A preference are analytical, rational, technical, logical, factual, and quantitative. B preference people are procedural, scheduled, conservative, organized, sequential, reliable, tactical, and administrative. C preference students are supportive, interpersonal, expressive, sensitive, symbolic, musical, and reaching out. D preference thinkers are more strategic, visual, imaginative, conceptual, and simultaneous.

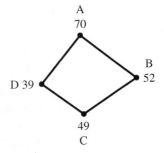

Figure 12.1   Sketch of student's thinking preference.

12.5 DISCUSSION

Apparently the student shown in the example has a strong preference for the A quadrant = analytical. Studies have shown that engineering faculty are predominantly A thinkers. Students change by virtue of practice to A thinkers by the end of their careers. Design students are generally D thinkers.

Everyone is creative and can learn to be more creative. Everyone can learn to use the D-quadrant thinking abilities of the brain more frequently and more effectively. So don't assume that creativity can't be learned and only a few have that ability. Another false assumption is the perception that an intelligent mind is a good thinker. According to Edward de Bono, highly intelligent people who are not properly trained may be poor thinkers for a number of reasons:

1. They can construct a rational, well-argued case for any point of view and thus do not see the need to explore alternatives.
2. Because verbal fluency is often mistaken for good thinking, they learn to substitute one for the other.
3. Their mental quickness leads them to jump to conclusions from only a few data points.
4. They mistake understanding with quick thinking and slowness with being dull witted.
5. The critical use of intelligence is usually more satisfying than the constructive use. Proving someone wrong gives instant superiority but doesn't lead to creative thinking.

12.6 PROBLEMS

1. Which quadrant would you consider a dominant quadrant for the following well-known figures?
   a. Newton
   b. Galileo
   c. Einstein
   d. Leonardo da Vinci
   e. Malcolm X
   f. Bach
   g. Martin Luther King, Jr.
   h. George Bush, Jr.
   i. Bill Clinton
   j. Ben Franklin
   k. Gandhi
2. Would you be able to change your thinking style? Why or why not?
3. If you are organizing a brainstorm session, which thinking styles would you like to have in the session and why?
4. Which role would you like to play in a team? Justify your answer based on your findings in the questionnaire.

13 Lab 2: Team Dynamics

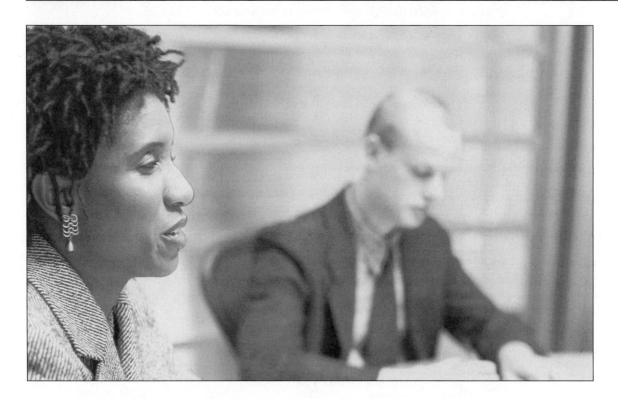

The design team is a group of individuals that must work together. Successful teams do not occur overnight. Effort and time need to be devoted to nurturing a successful team.

13.1   PURPOSE

Design as well as other major tasks requires individuals to work in teams. Major companies as well as small businesses recognize the need for employees to work in teams to achieve a goal. Managerial initiatives such as total quality management (TQM) and process reengineering recommend teams as the preferred way to organize and accomplish work. Concurrent engineering relies on cross-functional teams to enhance product development and innovation. Many undergraduate classes feature team assignments to help students develop the skills required for a successful career. However, especially in engineering, little class time is devoted to solving issues as a team. Thus the objectives of this lab are as follows:

1. Discuss the dynamics of team formation.
2. Present methodology for effective teams.

13.2   THEORY

A team is a collection of individuals whose work is interdependent and who are collectively responsible for accomplishing a performance outcome.[1] Thus a collection of individuals in a music consort doesn't constitute a team although all the individuals are in the same place at the same time for the same purpose. The keyword is *collectively*. A group of students assembled at the start of the semester doesn't compose a team unless the students are working collectively toward accomplishing an outcome.

Regardless of the performance outcome, all teams are characterized by the following features:[2]

1. A dynamic exchange of information and resources among team members.
2. Task activities coordinated among individuals in the group.
3. A high level of interdependence among team members.
4. Ongoing adjustments to both the team and individual task demands.
5. A shared authority and mutual accountability for performance.

Many students dread group projects. Past experience with poorly functional teams and a lack of school system training may have built this fear in students. Larson and LaFasto[3] studied the characteristics of effective functional teams. They reported that successful teams have the following characteristics:

---

[1]Mohrman, S. A., and Mohrman, A. M. *Designing and Leading Team-Based Organization: A Workbook for Organizational Self Design.* San Francisco: Jossey-Bass, 1997.
[2]McGoutry, J., and De Meuse, K. P. *The Team Developer.* New York: Wiley, 2001.
[3]Larson, C. E., and LaFasto, F. M. J. *Teamwork: What Must Go Right/ What Can Go Wrong.* Newbury Park, CA: Sage, 1989.

1. A clear, challenging goal; this goal gives the group members something to shoot for. The goal is understood and accepted by the entire group.
2. A result-driven structure; the roles of each member are clear, a set of accountability measure is defined, and an effective communication system is established.
3. Competent and talented team members.
4. Commitment; team members put the team goals ahead of individual needs.
5. Positive team culture. This factor consists of four elements: (1) honesty, (2) openness, (3) respect, and (4) consistency in performance.
6. Standard of excellence.
7. External support and recognition. Effective teams receive the necessary resources and encouragement from outside the group.
8. Effective leadership.

Design can be considered as a social activity in which a collective effort from a team is put together to produce the required output. Keep in mind that successful teams do not occur automatically or overnight. Effort and time need to be devoted to nurturing a successful team. Also, there are stages according to which teams evolve:

1. *Forming:* At this stage the group members still work as individuals; they do not contribute to the group as a whole but look out for themselves.
2. *Storming:* At this stage the group realizes that the task requires collective contribution and not much has been done. This prompts disagreements, blaming, and impatience with the process. Some members try to do it all on their own and avoid collaboration with team members.
3. *Norming:* When the team's objectives are worked out collectively, the common problems or goals begin to draw individuals together into a group, although the sense of individual responsibility is still very strong. Conversation among the team members helps direct efforts toward better teamwork and increases members' sense of responsibility for the team objective.
4. *Performing:* At this stage the team members have accepted each others' strengths and weakness and have defined workable team roles.

Team members with all four thinking styles compose a better team (see Lab 1). McGourty et al.[4] and DeMeuse and Erffmeyer[5] present four behaviors that are needed for effective team performance:

---

[4]McGourty, J., Tarshis, L. A., and Dominick, P. "Managing Innovation: Lessons from World Class Organizations," *International Journal of Technology Management*, Vol. 3, p. 4, 1996.
[5]De Meuse, K. P., and Erffmeyer, R. C. "The Relative Importance of Verbal and Nonverbal Communication in a Sales Situation: An Exploratory Study." *Journal of Marketing Management*, Vol. 4, pp 11–17, 1994.

1. *Communication team behavior:* Team members need to create an environment in which all members feel free to speak and listen attentively. The following behavioral practices for both roles must be followed:
   a. Listen attentively to others without interpreting.
   b. Convey interest in what others are saying.
   c. Provide others with constructive feedback.
   d. Restate what has been said to show understanding.
   e. Clarify what others have said to ensure understanding.
   f. Articulate ideas clearly and concisely.
   g. Use facts to get points across to others.
   h. Persuade others to adopt a particular point of view.
   i. Give compelling reasons for ideas.
   j. Win support from others.

2. *Decision-making team behavior:* Decision making is done by the team, not for the team. The following behavioral practices must be followed:
   a. Analyze problems from different points of view.
   b. Anticipate problems and develop contingency plans.
   c. Recognize the interrelationships among problems and issues.
   d. Review solutions from opposing perspectives.
   e. Apply logic in solving problems.
   f. Play a devil's advocate role when needed.
   g. Challenge the way things are being done.
   h. Solicit new ideas from others.
   i. Accept change.
   j. Discourage others from rushing to conclusions without facts.
   k. Organize information into meaningful categories.
   l. Bring information from outside sources to help make decisions.

3. *Collaboration team behavior:* Collaboration is the essence of team work; it involves working with others in a positive, cooperative, and constructive manner. The following behavioral practices must be followed:
   a. Acknowledge issues that the team needs to confront and resolve.
   b. Encourage ideas and opinions even when they are different from your own.
   c. Work toward solutions and compromises that are acceptable to all.
   d. Help reconcile differences of opinion.
   e. Accept criticism openly and nondefensively.
   f. Share credit for success with others.
   g. Cooperate with others.
   h. Share information with others.
   i. Reinforce the contribution of others.

4. *Self-management team behavior:* The following behavioral practices must be followed:
   a. Monitor progress to ensure that goals are met.
   b. Create action plans and timetables for work session goals.
   c. Define task priorities for work sessions.
   d. Stay focused on the task during meetings.

    e. Use meeting time efficiently.

    f. Review progress throughout work sessions.

    g. Clarify roles and responsibilities of others.

Each of these behaviors when followed will assure an effective team. It is important that distinct team roles be defined. The following roles apply to a team of four members:

1. Captain, who possesses behaviors and skills described in self-management team behavior.
2. Chief engineer, who possesses behaviors in decision-making team behavior.
3. Human resources person, who possesses behaviors described in collaboration team behavior.
4. Spokesperson, who possess behaviors described in communication team behavior.

Although this lab addressed some important points in team dynamics, there are many topics that need to be discussed. Some of these topics are addressed as part of the critical thinking activities. It is recommended that once students finish the exercise in the procedure section they devote time and effort to addressing the questions in the critical thinking section.

13.3   PROCEDURE

1. Arrange in teams of four members each. Make sure that all thinking styles are represented in the team as much as possible.
2. Map each of the team behaviors presented in Section 13.2 to a thinking style. Discuss your findings.
3. Once you have done the mapping, assign roles to each team member based on the recommendations in the previous section. Ask the human resources person to play the role of reflector, who will write a report describing the team dynamics during the brainstorming session. Assign the spokesperson to be a recorder of the actions and ideas generated during the brainstorming session. The chief engineer, once the session is over, will judge all the ideas and submit a report to the captain (CEO) of the team. The captain, once the exercise is over, will submit a report detailing the team activity to the board of directors (the classmates and instructor).
4. Consider the following for a brainstorming exercise:
   The energy crisis in California in the summer of 2001 has decision makers in your community concerned about a similar crisis. Brainstorm the following energy conservation topics:
   a. How can houses be built or improved to be more energy efficient?
   b. How can your local community transportation system be changed to be more energy efficient?

13.4 PROBLEMS

1. In his book titled *Human Behavior at Work* (McGraw-Hill, 1977), Keith Davis presented what he called "The Ten Commandments of Good Listening." List these ten commandments and discuss how you would use them to develop your feedback skills.
2. Thomas and Kilman, in *The Thomas–Kilman Conflict Mode Instrument* (Xicom, 1974), identified five general styles of conflict management. Summarize these styles.
3. Conducting successful meetings is not an easy task. Deborah Harrington-Mackin, in *The Team Building Tool Kit: Tips, Tactics and Rules for Effective Workplace Teams* (AMACOM, 1994), identified many symptoms that lead to unsuccessful team meetings. List 10 of these symptoms and discuss how you would overcome them.

Lab 3: Building Your Own Web Page

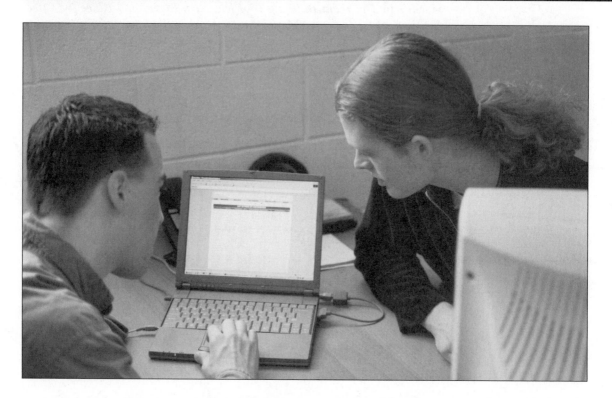

The Internet and e-mail have become the dominant communication tools of our time. This chapter will show you how to develop a Web page using Microsoft Office tools.

14.1　PURPOSE

The Internet and e-mail have become dominant communication modes. This lab shows how to develop a Web page using Microsoft Office tools. It is assumed that students have access to computer labs that are connected to a UNIX server. The Web server in this lab is an Apache server. Instructors are advised to direct students if the computer setup is different in their computer labs.

14.2　PROCEDURE

1. Create public_html directory
   a. Login to a PC in any computer lab.
   b. Double click the Telnet icon. A window will appear. Select connect. Drag the mouse down to remote system and release. A small window will show up. In the Host Name field, type (server name). Then click on the connect button.
   c. Login to your UNIX account by entering your login id and password.
   d. After you login you will see server_name:/home/your_login_ID>¿.
   e. Type mkdir public_html and then hit enter to create a new directory called public_html.
   f. To see if you have created the directory, type ls (list); you should see a directory name public_html listed.
   g. Go to the connect heading. Click and hold while you drag the mouse down to disconnect.
2. Create index.html file
   a. On your PC double click on the MSWord icon. If the Wizard appears, select start MSWord. If the User Name dialog box appears, select OK to start using MSWord.
   b. Type a welcoming statement, such as Welcome to (your_name) Homepage, Hello, etc. Center the text using the MSWord formatting tools.
   c. Save the document in your public_html directory. To do that while you are still in Word,
      i. Click on File menu and select Save As HTML. In the dialog box that appears, locate your home directory in the college system. Click on the arrow next to the Save In box. The home directory should be in the F drive.
      ii. Locate your public_html directory.
      iii. Name the file as index.html.
      iv. Make sure that HTML document appears in the Save As type box.
      v. Click on the save icon in the dialog box.

3. View your Web page.
   a. You can preview your Web page using MSWord by selecting Web Page Preview from the File menu. MSWord will open the browser for you to see your Web page.
   b. View your Web page using the browser.
   c. In the Netsite box on your Netscape browser, type *http://www.yourschoolservername.edu/(youruserID)*. This is the uniform resource locator (URL) address of your home page.

4. Modify your page: There are many features that you can add to your home page. In this lab a few features are introduced. The techniques used in these features can be applied for other features.
   a. Changing colors
      i. Go back to your index.html file and highlight the text you would like to change in color. Change the color to any color you may like using the icon that has the letter A underlined with a red line.
      ii. Save the file by clicking on the file menu and dragging the mouse to the Save icon.
      iii. Preview or review your Web page as in Step 3. You can use Alt + Tab to switch between MSWord and the browser.
   b. Changing font style and/or size
      i. Go back to your index.html file and highlight the text you want to reformat. Click on the Format heading and choose Font from the listed menu.
      ii. A dialog box will appear. Choose the style you like, font size, or color you wish to change to.
      iii. Save the file by clicking on the File menu and dragging the mouse to the Save icon.
      iv. Preview or review your Web page as in Step 3.
   c. Adding your e-mail to the Web page
      i. Add a line that leads people to your e-mail such as "Please e-mail me at (*youruser_ID*)@*yourschoolserver.edu*".
      ii. An e-mail link will be created to you.
      iii. Save the file by clicking on the File menu and dragging the mouse to the Save icon.
      iv. Preview or review your Web page as in Step 3.
   d. Creating a link
      i. Add text that you want to be linked to another Web page, such your school Web page.
      ii. Highlight the text and click on the Insert menu. Choose the Hyperlink icon. A dialog box will appear.
      iii. Type the link address in the box labeled as link to file or URL *http://www.yourschoolserver.edu/instructor-name* and click on the OK button.
      iv. It is convenient to view immediately the changes to your Web page using the browser commands. After you have saved your changes, click on the Reload button on the browser toolbar. To reload correctly, Netscape caching needs to be turned off. To turn off the

caching, click on the Netscape Edit menu and choose Preferences. A box will appear. Choose advanced; then choose Cache. Set both the Memory Cache and the Disk Cache sizes to zero and clear both caches. Then click OK.

e. Adding a horizontal line

   i. To insert a colored horizontal line, click on Insert; then choose the Horizontal Line icon.

   ii. Choose a line; then save the file and reload your Web page.

f. Adding an image

   i. Choose Insert and then choose Picture from the menu. Another menu will appear to give you options from where you can choose the image. For now choose from file.

   ii. Choose an image from the given pictures; then click on Insert.

   iii. If your image doesn't show up when you view it using the browser, then go back to index.html file, click on the View menu, and drag the mouse to HTML Source. Another window will appear that contains the HTML source for your file. The problem may occur because MS saves the name of the images in lowercase and creates a link with a capital first letter. Change the names of the images to lowercase and save the file. You can view the image now.

   iv. To position the picture in the right place, highlight the image; then go to the Format heading and drag the mouse to the Picture icon. A dialog box will appear. Select the position of the picture from the dialog box.

g. Adding bulleted and numbered lists

You can create lists by using the icons on the toolbar. To change the format of the bullets or numbers, choose the Format heading; then choose Bullets and Numbering. A box with options will appear; click on OK.

h. Creating tables

Select the Table icon from the standard toolbar. Choose the size of table you want. The initial size is limited to 4 × 4. This can be adjusted by cut and paste later. To add columns, rows, and borders, choose the Table menu; then select the icon you need.

Tip: To avoid misalignment between your source file in MSWord and the way the document appears in the browser, try to avoid tabs or empty spaces in your text. Use the alignment icons (left, center, right) in the formatting toolbar instead.

Lab 4: MS Project 2000® as a Tool for Project Management

Project management and scheduling involves sequencing and allotting time to all project activities. MS Project 2000® is a software program designed to assist the project manager in tracking the project's activities. Tools such as this calendar help the project manager review activities and dates at a glance.

15.1   PURPOSE

In this lab you will be introduced to some main concepts of project management and some techniques to control and monitor a project using Microsoft Project 2000 software. This lab was developed by Adnan Bashir (bashir@eng.fsu.edu).

15.2   INTRODUCTION

Management is the process of getting things done through the effort of other people by planning, organizing, and controlling these processes.

The project is a complex effort usually less than three years in duration, made up of interrelated tasks performed by various organizations within a well-defined objective, schedule, and budget. The objective of project management is to achieve proper control of the project to assure its completion on schedule and within budget, achieving the desired quality of the resulting product or services.

The reasons for project planning are as follows:

- Eliminate or reduce uncertainty.
- Improve efficiency of the operation.
- Obtain a better understanding of the project objectives.
- Provide a basis for monitoring and controlling work.

If the task is not well understood, then during task execution more knowledge is gleaned, which leads to changes in resource allocation schedules and properties. If the task is well understood prior to being performed, then work can be preplanned.

15.3   PROJECT SCHEDULING

Project scheduling involves sequencing and allotting time to all project activities. Managers decide how long each activity will take and compute the resources (personnel, equipment, and materials) needed for each activity. One popular project scheduling approach is the Gantt chart. **Gantt charts** are planning charts used to schedule resources and allocate time. Gantt charts are an example of a widely used, nonmathematical technique that is very popular with managers because it is so simple and visual. The major discrepancy with Gantt charts is the inability to show the interdependencies between events and activities. Interdependencies are shown through the construction of networks. Network analysis can provide valuable information for planning, scheduling, and resource management. The program evaluation and review technique (PERT) and critical path method (CPM) are very popular and widely used network techniques for large and complex projects. **PERT** is a technique to enable managers to schedule, monitor, and control large and complex projects by employing three time estimates for each activity. **CPM** is a network technique using only one time factor per activity that enables managers to schedule, monitor, and control large and complex projects. PERT and CPM were developed in 1958 and 1959, respectively, and spread rapidly throughout all industries. CPM was concentrated in the construction and process industries.

15.3.1 Framework of PERT and CPM

- Define the project and prepare WBS.
- Develop the relationship between the activities.
- Draw the network connecting all the activities.
- Assign a time/cost estimate to each activity.
- Compute the longest path through the network (critical path).
- Use the network to help plan, schedule, monitor, and control the project.

Computerized PERT/CPM reports and charts are widely available today on personal computers. Some popular software are Primavera (by Primavera Systems, Inc.), MS Project (by Microsoft Corp.), MacProject (by Apple Computer Corp.), Pertmaster (by Westminster Software, Inc.), VisiSchedule (by Paladin Software Corp.), and Time Line (by Symantec Corp.). In this lab we will use MS Project 2000.

15.4 USING MS PROJECT 2000

The first step to start the project is to build the project calendar that will be used in the project. Figure 15.1 shows the standard calendar, which is the default in MS Project 2000.

Figure 15.1 Standard project calendar.

Figure 15.2   Creating a new calendar.

You can create a new calendar to suit your project working days, vacations, number of shifts, and so on. To create a new calendar, use the following steps:

- Go to Tools.
- Go to Change Working Time.
- Select New.
- Insert the new name of the calendar (proj1).
- Choose Copy from Standard.

You will get a new calendar called proj1. You can easily change the nonworking days to Thursday and Friday by assigning them to non-working time, and assign the Saturday and Sunday to default (working days). Also, we can increase or decrease the number of working hours in a day by using nondefault working time and fill in for propitiate working hours. Figure 15.2 shows the new calendar (proj1).

Once you set up the calendar to conform to your project, you can easily enter the tasks and the duration of your project, as shown in Figure 15.3.

Once the task name and duration are entered, you can insert the relationship among the tasks by inserting a precedence column and then entering the task ID to identify the task. Figure 15.4 shows the task, duration, and the relationship among the tasks. The first activity shows a task summary with duration covering the subtasks. The arrows between the bars represent the relationship among the tasks; in this case the relationship is Finish to Start (FS), which means that the activity can not start until the preceding activity is completed.

Figure 15.4 shows the Gantt chart (scaled time bar chart for each task), task name, task duration, start, finish, and precedence. If you do not specify the start date of the project, MS Project 2000 will consider the current date as the project start date (as default). This Gantt chart is used for most projects because it is simple and easy to understand, and it is easy to extract summary information for the project. You can add any information to appear in the Gantt chart, such as resources, float (slack), percentage completion, actual start, actual finish, late start,

Figure 15.3 Task name and duration.

Figure 15.4 Tasks with their duration and relationships.

late finish, actual cost, baseline cost, milestones, remaining duration, and remaining cost. All of these can be added as follows:

- Go to Insert.
- Select Column.
- Pick the type of information you want to add.
- Press OK.

If you double click on any task, a new window will appear that contains the following subwindows:

- General: Gives the name, duration, the completion, start and finish dates of that task.
- Predecessors: All the predecessors of this task will appear under this window showing their ID, name, type, and the lag ID if available.
- Resources: You can assign the resources required to complete the task by entering the resource name and how many units needed.
- Advanced: This gives you the ability to insert time constraints, the task type, and the calendar. It also shows the work break structure of the task (WBC).

Figure 15.5 shows the task information window for task number 2.

To enter the resources of the project, go to resource sheet, and start to fill in that sheet as shown in Figure 15.6. Then assign the resources to each task as shown in Figure 15.7:

- Go to tools.
- Go to Resources, Assign resources.
- Pick the resources and units for that task.

MS Project 2000 provides several predefined reports, as shown in Figure 15.8. There is an option to customize any report you like. For example,

- Go to View, click Reports, and click Custom.
- Click select button; in the report box select Task, and then copy button.
- In the name box, type Custom Task Report.

Figure 15.5   Task information window.

Figure 15.6   Resource sheet.

Figure 15.7   Assign task resources.

Figure 15.8   Types of reports in MS Project 2000.

Figure 15.9   Customized task report.

- In the period box, select Months from the drop-down list.
- In the table box, select Summary from the drop-down list.
- Press OK (Figure 15.9).
- Click OK to close the Task Report dialog box.
- In the Custom Reports dialog box, click preview.
- MS Project applies the custom report settings you choose, and the report appears.
- The print preview window appears (Figure 15.10).

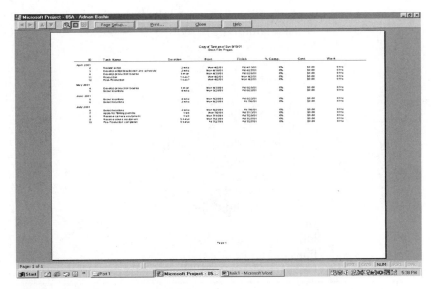

Figure 15.10   Customized summary task report.

The custom report in Figure 15.10 shows the fields displayed on the Summary Task table but divides the tasks by month.

15.5 USING MS PROJECT — SAMPLE

Consider the following list of activities for building a small house project:

1. Determine the ES, LS, EF, and LF.
2. Determine the total float.
3. Determine the critical path and the total project duration.

15.5.1 SOLUTION

- For this example we will keep the default calendar (Standard).
- The project will start on 10/15/01. To enter this information,
  a. Go to project
  b. Select Project Information, and enter the start date of the project.
- Enter the task name, duration, and predecessor for each task in MS Project software (see Table 15.1).

TABLE 15.1   Example

Activity	Description	Predecessor	Duration (Days)
A	Clear site	—	1
B	Bring utilities to site	—	2
C	Excavate	A	1
D	Pour foundation	C	2
E	Outside plumbing	B, C	6
F	Frame house	D	10
G	Electric wiring	F	3
H	Lay floor	G	1
I	Lay roof	F	1
J	Inside plumbing	E, H	5
K	Shining	I	2
L	Outside sheathing insulation	F, J	1
M	Install windows/outside doors	F	2
N	Brick work	L, M	4
O	Install walls and ceiling	G, J	2
P	Cover walls and ceiling	O	2
Q	Install roof	I, P	1
R	Finish interior	P	7
S	Finish exterior	I, N	7
T	Landscape	S	3

- Put the grid line for the major and minor columns as follows:
    a. Click the right button of the mouse while the cursor is in the bar chart area.
    b. Select the grid option.
    c. Select the major column and choose the suitable type of line and its color.
    d. The same for the minor column.
- In order to get a full view of the Gantt chart,
    a. Go to View, and select Zoom.
    b. Select the option Entire Project.

The outputs of these steps are shown in Figure 15.11.

- In order to find the ES, EF, LS, LF, and the total float,
    a. Go to Insert.
    b. Select Column.
    c. From the field name, select Early Start, and press OK.
    d. Repeat the same for EF, LS, LF, and total float.
- To show the critical path and the critical activities,
    a. Go to Format.
    b. Select Bar Style.

Figure 15.11   Gantt chart for the example.

Figure 15.12   Gantt chart with ES, EF, LS, LF, and total slack.

    c. Click Insert Row, and type Critical Activity.
    d. Select the shape of the bar and its color (red).
    e. In the field shown for task, select Critical, and press OK.
    f. Note that the critical tasks should have total slack = 0.

The outputs are shown in Figure 15.12.

- The total project duration can be obtained as follows:
    a. Go to Project.
    b. Select Project Information.
    c. Select Statistics. Information about the project will be displayed as shown in Figure 15.13.
- To show the network diagram,
    a. Go to View.
    b. Select Network Diagram.
    c. Use zoom in and zoom out to see the desired view.

Figure 15.14 shows the network diagram for the preceding example, as well as the critical path on the network with the red color. Each node (box) contains task name, duration, start date, finish date, and resources. Each node is connected to another node according to the relationship among the tasks.

Figure 15.13 Project information statistics.

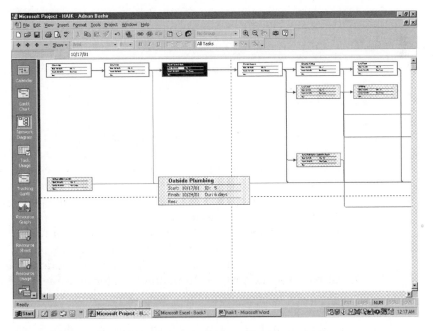

Figure 15.14 Part of the network diagram for the example.

15.6 PROBLEMS

1. What are the differences among Gantt, PERT, and CPM?
2. What is the dummy activity and why do we use it?
3. Construct a project network consisting of 12 activities (from 10 to 120), with the following relationships:
   a. 10, 20, and 30 are the first activities and are begun simultaneously.
   b. 10 and 20 must precede 40.
   c. 20 precedes 50 and 60.
   d. 30 and 60 must precede 70.
   e. 50 must precede 80 and 100.
   f. 40 and 80 must precede 90.
   g. 70 must precede 110 and 120.
   h. 90 must precede 120.
   i. 100, 110, and 120 are terminal activities.
4. Draw the CPM network associated with the activities listed in Table 15.2 for a certain project. How long will it take to complete the project? What are the critical activities?

TABLE 15.2   Table for Problem 4

Activity	Predecessor(s)	Time(min.)
A	—	2
B	A	4
C	A	6
D	B	6
E	B	4
F	C	4
G	D	6
H	E, F	4

5. For problem 3 find the ES, EF, LS, LF, and total float for each activity.
6. ABC company has a small project with the activities listed in Table 15.3. The manager of the project has been very concerned with the amount of time it takes to complete the project. Some of his workers are very unreliable. A list of activities with their optimal completion time, most likely completion time, and undesirable completion time (all in days) are given in Table 15.3. Determine the expected completion time and the variance for each activity, and draw the PERT network diagram.
7. In Problem 6, determine the total project completion time and the critical path for the project. In addition, determine the ES, EF, LS, LF, and total slack for each activity.
8. In Problem 6, by using MS Project, produce a report that shows a list of the tasks that have an expected duration more than 5 days.

TABLE 15.3		Table for Problem 6		
Activity	**a**	**m**	**b**	**Predecessor(s)**
A	3	6	8	—
B	2	4	4	—
C	1	2	3	—
D	6	7	8	C
E	2	4	6	B, D
F	6	10	14	A, E
G	1	2	4	A, E
H	3	6	9	F
I	10	11	12	G
J	14	16	20	C
K	2	8	10	H, I

9. Redo Problem 6, using the following calendar:
   a. Working hours per day is 10 hours.
   b. Working days per week is 6 days (Sunday is a day off.)

15.7 SELECTED BIBLIOGRAPHY

1. Kerzner, H. *Project Management: A Systems Approach to Planning, Scheduling, and Controlling*, 7th Ed., New York: Wiley, 2001.
2. Heizer, J. and Render, B. *Principles of Operations Management*, 4th Ed., Upper Saddle River, NJ: Prentice Hall, 2001.
3. Chatfield, C. and Johnson, T. *Step by Step Microsoft Project 2000*. Seattle, WA: Microsoft Press, 2000.

16 Lab 5: Presentation Style

Even the best idea or design will not be effective if you cannot properly communicate your ideas to others. This chapter provides startup training in presentation styles.

16.1 PURPOSE

Unlike the advertising executive, the engineer is ill equipped to sell his or her ideas. Second-hand information represented by company officials may not answer all the client's questions. This lab provides startup training for your presentation style.

16.2 ESTABLISH OBJECTIVES

Every presentation should have an objective. The speaker's main objective is to deliver the message (objective) to the audience. The objectives may vary from one speech to another. To identify the real objective, ask the following question: "If everything goes perfectly, what do I intend to achieve?" Also, know your audience and their education level. In most cases, time is limited. It is of utmost importance to keep within the scheduled time for the question/answer session. Doing so requires detailed planning. Different audiovisual tools can be used in the presentation, such as slides, models, transparencies, and the Web. If more than one medium is used, allow for time to shift from one medium to another. One easy way to gauge the effectiveness of a presentation is for one team member to play the role of the speaker and the rest of the team to pose as the audience (a team member may also play the devil's advocate).

16.3 ORAL PRESENTATION OBSTACLES

To sell your ideas to others, you should first be convinced that your ideas will accomplish the task at hand. Oral presentation requires a high degree of creative ability. Humans resist change; they like familiar methods. Changes require additional effort, which humans in general resist. The following are typical reactions to proposed changes:

- We tried before.
- It is too radical a change.
- We have never done that before.
- Get back to reality.
- We have always done it this way.
- I don't like the idea.

Keep in mind that

- Only 70% of spoken words are actually received and understood. Complete understanding can come through repetition and redundancy in speech.
- People best understand three-dimensional objects. Two-dimensional projections need to be transmitted with added details.
- When considering a proposed change, people usually perceive potential problems.
- Convey ideas so that they are interpreted with the least expenditure of energy.

16.4   ORAL PRESENTATION DO'S AND DON'TS

1. Know your audience thoroughly.
2. Never read directly from notes, a sheet, or directly from an overhead projector.
3. Bring your audience up to speed in the first few moments.
4. Stay within the time allotted.
5. Include relevant humorous stories, anecdotes, or jokes (only if you are good at it).
6. Avoid using specialized technical jargon. Explain the terms if you feel the audience may not know them.
7. State your message clearly.
8. Practice, practice, practice. You may want to memorize the introduction and concluding remarks.
9. The dry run is a dress rehearsal. Use it to iron out problems in delivery, organization, and timing.
10. Avoid mannerisms: Speak confidently but not aggressively.
11. Maintain eye contact with some audience members and keep shifting that contact throughout the presentation.
12. Never talk to the board or to empty space.
13. Present the material in a clever fashion, but not in a cheap and sensational fashion. Be genuinely sincere and professional.
14. Logical presentation is much more critical in oral than in written presentation.

16.5 ORAL PRESENTATION TECHNIQUES

1. Visual aids (sketches, graphs, drawings, photos, models, slides, transparencies, the Web) often convey information efficiently and effectively. Visual aids permit a dual sense of hearing and seeing, and they help the speaker.

2. Limit slides to not more than one per minute.

3. Each slide should contain one idea.

4. The first slide should show the title of your presentation and names of collaborators.

5. The second slide should give a brief outline of the presentation.

6. The last slide should summarize the message you just delivered.

7. If you need to show a slide more than once, use a second copy.

8. Avoid leaving a slide on the screen if you have finished discussion on that topic.

9. Never read directly from the slide. Spoken words should complement the slides. Prepare notes for each slide and use them during practice.

10. Use graphs to explain variations. Clearly label the axis, data, and title. Acknowledge the source.

11. Every graph should have a message (idea). Color should enhance the communication, not distract.

12. Audiences respond to well-organized information. That includes
    (a) Efficient presentation
    (b) All assumptions clearly stated and justified
    (c) Sources of information and facts clearly outlined

13. Begin with the presentation of the problem and conclusion/recommendation (primary goal).

14. Finish ahead of time and be prepared for the question/answer session.

16.6   QUESTION/ANSWER SESSION

The question/answer session is very important. It shows the audience's enthusiasm, interest, and attention. In the Q/A session you should

1. Allow the questioner to complete the question before answering.
2. Avoid being argumentative.
3. Do not let the questioner feel that the question is stupid.
4. Adjourn the meeting if the questions slack off.
5. Thank the audience one final time after the Q/A session.

16.7   PROBLEMS

Choose one of the following topics and prepare a 5-minute presentation for your classmate or teammates. Adhere to the instructions provided in this lab. Your classmates or teammates will provide anonymous evaluation scores through the instructor. The evaluation scores consists of the following elements:

1. Level of preparation (out of 10 points). What level of preparation was expressed in the slides' quality and content?
2. Level of audience engagement (out of 5 points). Did the speaker maintain eye contact with the audience?
3. Level of professionalism (out of 5 points). Was the speaker dressed properly, and did he or she act professionally during the presentation?
4. Communication skills (out of 5 points). Was the presentation clear? Was the speaker able to convey the message?
5. Time management (out of 5 points). Did the speaker leave time for questions? How much time is really needed to convey the message?
6. Overall performance (out of 10 points).

The instructor will collecte the evaluations and pass them on to the presenter. The purpose of this exercise is to provide you with an evaluation of your presentation skills. You need to enhance the strong aspects and work on improving the weak aspects of presentation technique.

The following is a suggested list of topics. Students may choose a different topic with instructor consent.

1. How can a person maintain reading time in a busy schedule?
2. How can a school library raise funds for more resources?
3. How can engineers keep up with the latest technology?
4. How can you maintain a steady study schedule when you have tons of homework?

5. What are the pros and cons of teamwork?
6. How can teamwork become a nightmare?
7. How can teamwork become fruitful?
8. Why is design a social activity?
9. Compare the old versus the new paradigms of design.

17 Lab 6: Ethics

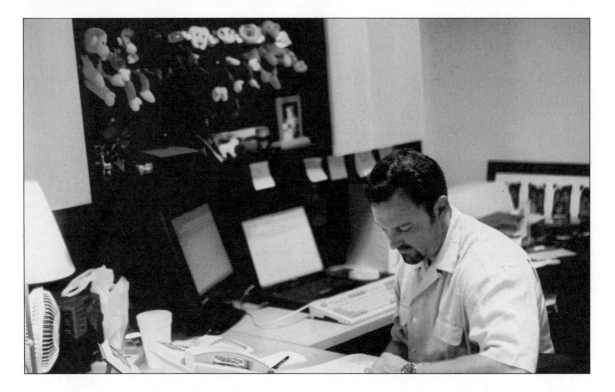

Ethics is a part of all professional careers, but play an extremely important role in engineering. In this chapter, we will discuss the importance of engineering ethics using two case studies.

17.1 PURPOSE

Ethics is a part of all professional careers. This lab introduces case scenarios that deal with professional ethics. Before you start this lab you will need too visit the following Web pages and familiarize yourself with their content:

1. http://web.mit.edu/ethics/www/essays/probcase.html by Caroline Whitebeck.
2. http://www.cwru.edu/affil/wwwethics/. After you review the content of this page, click on Problems. This will lead you to different case studies in engineering and science ethics.
3. http://www.onlinethics.org

17.2 PROCEDURE

You will need to submit two reports in the course of this lab:

1. *Report I assignment:* In this report you should (as a group)
   a. List the major points that were discussed by the Caroline Whitebeck paper.
   b. Discuss the paper and point out where you agree or disagree.
   c. Discuss your scenario and report your findings and arguments (The instructor will assign an ethical scenario to your team.)
2. *Report II assignment*
   a. Identify experts from the university and local community.
   b. Interview at least three experts and report your interview discussion.
   c. Analyze the three interviews.
   d. Make a recommendation/final argument based on your opinion and the interviews.
   e. Create a Web page for your report.

The following are the different scenarios that were obtained from the Internet. (One of these pages is from Case Western Reserve University, which is not reachable at the time of this writing.) The discussion at end of each scenario will help you get started; do should not limit your scope to those questions and discussion points when you report.

Scenario I

You are an engineer charged with performing safety testing and obtaining appropriate regulatory agency or outside testing laboratory ("agency") approvals of your company's product. The Gee-Whiz Mark 2 (GWM2) has been tested and found compliant with both voluntary and mandatory safety standards in North America and Europe.

Because of a purchase-order error and subsequent oversights in manufacture, 25,000 units of GWM2 ("bad units") were built that are not compliant with any of the North American or European safety standards. A user would be much more vulnerable to electric shock from a bad unit than from a compliant unit. Under some plausible combinations of events the bad unit user could be electrocuted. Retrofitting these products to make them compliant is not feasible because the rework costs would exceed the profit margin by far. All agree that because of this defect the agency safety labels will not be attached to the bad units, as per the requirements of the several agencies. Only two options exist:

1. Scrap the units and take the loss.
2. Sell the units.

An employee of the company notes that many countries have no safety standards of any kind for this type of product. It is suggested that the bad units be marketed in these countries. It is pointed out that many of these nations have no electrical wiring codes; or if codes exist they are not enforced. The argument is thus advanced that the bad GWM2 units are no worse than the modus operandi of the electrical practice of these countries. Assume that no treaties or export regulations would be violated in marketing the bad units to these countries.

Discussion

1. What is your recommendation?
2. Suppose one of the countries under consideration was the country of origin for you or your recent ancestors. Would this affect your recommendation?
3. Now suppose you are not asked for a recommendation, only an opinion. What is your response?
4. Suppose it is suggested that the bad units be sold to a third party, who would very likely sell the units to these countries. What is your comment?
5. You are offered gratis one of the bad units for your use at home, provided that you sign a release indicating your awareness of the condition of the unit, and that it is given to you as a test unit. (Assume that you can't retrofit it, and that the product could be very useful to you.) Would you accept the offer?
6. Suppose it is suggested that the offer in item 5 be made to all employees of the company. Your comment?

Scenario II

The U.S. Federal Communications Commission (FCC) Rule Part 15J applies to virtually every digital device (with a few exceptions) manufactured in the United States. The manufacturer must test and certify that the equipment does not exceed FCC-mandated limits for the generation of communications interference caused by conducted and radiated emissions. The certification consists of a report sent to the FCC for review. It is largely an honor system because the FCC has only a small staff to review an enormous number of applications. The FCC then issues a label ID, to be attached to each unit, that authorizes marketing of the product. Prior to receiving the label the manufacturer cannot offer for sale or advertise the product.

An EMC (electromagnetic compatibility) consultant operating a test site installs a new antenna system and finds that it results in E-field measurements consistently higher than those obtained with the old antennas. Both track within the site-calibration limits, and both antenna vendors claim National Institute of Standards and Technology (formerly National Bureau of Standards) traceability. Which system is the better in absolute calibration is thus unknown. There is not enough time to resolve this discrepancy before a client's new product must be tested for FCC Rules Part 15J compliance.

Discussion

1. Which antenna system should be used to test the product?
2. Is averaging the results ethical assuming that engineering judgment indicates that this procedure is valid?
3. Suppose the site was never properly (scientifically or statistically) calibrated. Should this fact be made known voluntarily to the FCC?

Scenario III

You are an engineer working in a manufacturing facility that uses toxic chemicals in processing. Your job has nothing to do with the use and control of these materials.

The chemical MegaX is used at the site. Recent stories in the news have reported alleged immediate and long-term human genetic hazards from inhalation or other contact with the chemical. The news items are based on findings from laboratory experiments, done on mice, by a graduate student at a well-respected university physiology department. Other scientists have neither confirmed nor refuted the experimental findings. Federal and local governments have not made official pronouncements on the subject. Several employee-friends have approached you on the subject and asked you to do something to eliminate the use of MegaX at your factory. You mention this concern to your manager, who tells you, "Don't worry, we have an industrial safety specialist who handles that."

Two months pass and MegaX is still used in the factory. The controversy in the press continues, but there is no further scientific evidence pro or con in the matter. The use of the chemical in your plant has increased, and now more workers are exposed daily to the substance than was the case two months ago.

Discussion

1. What, if anything, do you do?
2. Suppose you again mention the matter to your manager and are told, "Forget it, it's not your job." What now?
3. Your sister works with the chemical. What is your advice to her?
4. Your pregnant sister works with the chemical. What is your advice to her?
5. The company announces a voluntary phasing out of the chemical over the next two years. What is your reaction to this?

6. A person representing a local political activist group approaches you and asks you to make available to them company information regarding the amounts of MegaX in use at the factory and the conditions of use. Do you comply? Why or why not?

Scenario IV

The Zilch Materials Corporation employs you as a test engineer. The company recently introduced a new two-component composition-resin casting material, Megazilch, which is believed to have been well tested by the company and a few selected potential customers. All test results prior to committing to production indicated that the material meets all published specifications and is superior in performance, and lower in estimated cost, than competitors' materials used in the same kinds of applications.

Potential and committed applications for Megazilch include such diverse products as infants' toys, office equipment parts, interior furnishings of commercial aircraft, and the case material for many electronic products. Marketing estimates predict a 25% increase in the corporation's revenues in the first year after the product is shipped in production quantities.

The product is already in production, and many shipments have been made when you discover, to your horror, that under some conditions of storage temperature and other (as yet) unknown factors, the shelf life of the product is seriously degraded. In particular, it will no longer meet specifications for flame retardation if stored for more than 60 days before mixing, instead of the 24 months stated in the published specifications. Its tensile and compressive strength is reduced significantly as well.

Substantial quantities have been shipped, and the age and temperature history of the lots shipped is not traceable. To recall these would involve great financial loss and embarrassment to the company, and at this point it is not clear that the shelf life can be improved. Only you and a subordinate, a competent test technician, know of the problem.

Assume that no quick fixes by chemical or physical means are possible, and that the problem is real. That is, there are no mistakes in the scientific findings.

Discussion

1. What is the first action you would take relevant to this matter?
2. Suppose you express concern to your immediate supervisor, who tells you, "Forget it! It's no big deal, and we can correct it later. Let me handle this."
3. Suppose further that you detect no action after several weeks have passed since you told your supervisor. What now?
4. In item 3, assume you speak to your supervisor, who then tells you, "I spoke to the executive staff about it and they concur. We'll keep shipping product and work hard to fix it. We've already taken out all the stops, people are working very hard to correct the problem." What, if anything, do you do?
5. It is now three months since you told your supervisor, and in a test of product sampled from current shipments, you see that no fix has been incorporated. What now?

Scenario V

Marsha is employed as the City Engineer by the city of Oz, which has requested bids for equipment to be installed in a public facility. Oz is bound by law to purchase the lowest bid that meets the procurement specifications except "for cause."

The low bidder, by a very narrow margin, is Diogenes Industries, a local company. The Diogenes proposal meets the specifications. Marsha recommends purchase of the equipment from Diogenes.

After the equipment is installed, it is discovered that John, the Chief Engineer for Diogenes, is the spouse of Marsha. John was the engineer who had charge of the proposal to Oz, including the final authority on setting the price. As a result of this, Marsha is requested to resign her position for breach of the public trust.

Discussion

1. Was the city justified in seeking Marsha's termination of employment?
2. Suppose Marsha had never been asked to sign a conflict of interest statement. Would this affect your response to question 1?
3. Given the conditions of question 2, suppose Marsha had mentioned, before going to bid, in casual conversation with other persons involved in the procurement that she was married to the Chief Engineer at Diogenes. Does this affect your response?
4. Suppose Marsha and John were not married but shared a household. Does this affect your response?
5. Now suppose Marsha had made known officially her relation to John and the potential for conflict of interest before soliciting bids. Then suppose Marsha rejects the Diogenes bid because she is concerned about the appearance of conflict of interest. She then recommends purchase of the next lowest bid, which meets the specifications. Comment on Marsha's action.

18 Lab 7: Material Selection

Material selection is an important part of any design. Choosing the right material can either make or break the design. This chapter will discuss a few material properties and analytical tools to aid you in making the proper material selection.

18.1 PURPOSE

Choosing the right material in designing components is an important aspect of design. This lab introduces a few material properties and analytical tools with which to choose a material. The level of presentation is at an introductory level.

18.2 THEORY

18.2.1 Density

Density is one of the most important material properties because it determines the weight of the component. The range of material densities is based on the atomic mass and the volume the material occupies. Metals are dense because they consist of heavy and closely packed atoms. Polymers are formed from light atoms (carbons and hydrogen) and are loosely packed.

18.2.2 Melting Point

The melting point of a material is directly proportional to the bond energy of its atoms. Melting point becomes an important factor when the operating conditions of the material are in high-temperature environments; as in internal combustion engines and boilers. In design, the rule that is generally used is that the environment temperature should be 30% lower than the melting point of the material ($T/T_m < 0.3$). Designers should also keep in mind that when a material is subjected to high temperature and undergoes stress, creep occurs (slow change of dimensions over time).

18.2.3 Coefficient of Linear Thermal Expansion

The coefficient of linear expansion is generally defined as

$$\alpha = \frac{dL}{L \, dT}$$

where L is the linear dimension of the object, dL is the change in the linear dimension of the object, and dT is the change in temperature that cause the change in length.

18.2.4 Thermal Conductivity

The thermal conductivity constant is a property of the material by which one can determine whether the material is a good heat conductor or insulator. In metals, electrons are the carriers of heat. A relation between the electrical conductivity and heat conductivity is established as

$$\frac{k}{\sigma T} = 5.5 \times 10^{-9} \frac{\text{calohm}}{\text{sec } K^2}$$

where k is the thermal conductivity, s is the electrical conductivity, and T is the temperature in Kelvin.

18.2.5 Strength of Material

The strength of the material determines the amount of load a material can sustain before it breaks. In general, the design criterion is based on the yield stress of the material. The strength of the material is measured experimentally using a tensile testing machine. In a design-testing machine a specimen is pulled with a certain load and the elongation that is associated with the load is measured. The load is converted to stress and the elongation is converted into strain. There are two general representations of the data: the stress–strain diagram and the true-stress–true-strain diagram. In the stress–strain diagram the stress is measured by

$$s = \frac{\text{load}}{\text{initial area}} = \frac{P}{A_o}$$

and the strain by

$$e = \frac{\text{change in length}}{\text{initial length}} = \frac{\Delta L}{L_o}$$

In the true-stress–true-strain diagram the change in area after each loading is recognized and the true stress and true strains are obtained by

$$\sigma = \frac{\text{load}}{\text{instantaneous area}} = \frac{P}{A_{\text{inst}}}$$

$$E = \int_{L_o}^{L} \frac{dL}{L} = \ln\frac{L}{L_o}$$

The instantaneous area can be obtained by conservation of volume as

$$A_o L_o = AL = \text{constant}$$

Once the diagram is obtained, the modulus of elasticity can be obtained by finding the slope of the straight line before the yield tress.

Designers apply a safety factor based on the application and strength of the material.

18.2.6 Ductility

Ductility is an important property that can be used to determine if a material can be formed. It determines the capacity of the material to deform before it breaks. A material that cannot sustain deformation is called a brittle material. Ductility is measured as either percent elongation, $\%e$, in the material or percent reduction in area, $\%AR$:

$$e = \frac{L_f - L_o}{L_o} \times 100$$

or

$$AR = \frac{A_o - A_f}{A_o} \times 100$$

18.2.7 Fatigue Properties

When the component is subjected to cyclic loading, fatigue properties become essential. In this case the failure will be introduced at stress lower than the yield stress. A correlation that has been used to estimate the fatigue limit is

$$\text{Fatigue limit} = 0.5 S_{\text{uts}}$$

where S_{uts} is the ultimate tensile strength.

18.2.8 Impact Properties

The impact property of a material is its resistance to fracture under sudden impact.

18.2.9 Hardness

Hardness is the measure of the material's resistance to indentation. Hardness can be measured in three different tests: Brinell (uses balls to indent in the surface), Vickers (uses pyramids to indent the surface), and Rockwell (measures the depth of indentation). The Brinell hardness number (BHN) can be obtained as

$$BHN = \frac{2P}{\pi D\left[D - \sqrt{D^2 - d^2}\right]}$$

where P is the load, D is the diameter of the indenter, and d is the measure of the diameter of indentation. The BHN is used to estimate the ultimate tensile strength $\left(S_{uts} = \dfrac{P_{max}}{A_o}\right)$ using

$$S_{uts} = 3.45BHN$$

in MPa

$$S_{uts} = 0.5BHN$$

in Ksi, where K is kips.

18.3 PROBLEMS

1. A tensile specimen was machined with a 25-mm diameter in the gauge section and was tested with the following load elongation:

	Load (N)	Gauge length (mm)
	0	50.000
	50,000	50.0613
	100,000	50.1227
Results	150,000	50.1848
	175,000	50.50
	200,000	51.35
	225,000	52.90
	231,000	53.40

The specimen has break load at 231,000 N (i.e., it breaks when you start adding more load after this load.) After fracture, the gauge length was measured to be 53.1 mm and the diameter was 23.25 mm.

a. Convert the data to engineering stress–strain and true-stress–true-strain.

b. Plot the engineering stress–strain and true-stress–true-strain.

c. Determine the modulus of elasticity (Young's modulus).

d. Determine the fracture strength, which is also called the tensile strength.

e. Determine the ductility in %*e* and %*AR*.

f. Plot the *ln*–*ln* curve of the true-stress and true-strain (i.e., *ln* true-stress versus *ln* true-strain) and find the slope of the line.

2. A steel exhibits a 500 BHN. Use the empirical relations to estimate the fatigue limit of the steel. If a safety factor of 2 is applied, what would be this limit?

19 Lab 8: Ergonomics

Ergonomics is defined as "the study of the problems of people adjusting to their work environment." Engineers and designers must keep the effects on human users forefront in their product designs.

19.1 PURPOSE

This lab provides an introduction to some of the ergonomics issues in design.

19.2 WORKED EXAMPLE

Table 19.1 shows a series of established standards.[1] The following Web page shows detailed drawings of the dimensions: http://www.openerg.com. Refer to Table 19.1 for this example.[2]

TABLE 19.1 Anthropometric measured data in mm of U.S. adults, 19 to 60 years of age. According to Gordon, Churchill, Clauser, et al. (1989), who used the numbers in brackets

Dimension	Men				Women			
	5th %ile	Mean	95th %ile	SD	5th %ile	Mean	95th %ile	SD
1. Stature	1647	1756	1867	67	1528	1629	1737	64
2. Eye height, standing [D19]	1528	1634	1743	66	1415	1516	1621	63
3. Shoulder height (acromion), standing [2]	1342	1443	1546	62	1241	1334	1432	58
4. Elbow height, standing [D16]	995	1073	1153	48	926	998	1074	45
5. Hip height (trochanter) [107]	853	928	1009	48	789	862	938	45
6. Knuckle height, standing	Na	Na	Na	Na	Na	Na	Na	Na
7. Finger height, standing [D13]	591	653	716	40	551	610	670	36
8. Sitting height [93]	855	914	972	36	795	852	910	35
9. Sitting eye height [49]	735	792	848	34	685	739	794	33
10. Sitting shoulder height (acromion) [3]	549	598	646	30	509	556	604	29
11. Sitting elbow height [48]	184	231	274	27	176	221	264	27
12. Sitting thigh height (clearance) [104]	149	168	190	13	140	160	180	12
13. Sitting knee height [73]	514	559	606	28	474	515	560	26
14. Sitting popliteal height [86]	395	434	476	25	351	389	429	24
15. Shoulder-elbow length [91]	340	369	399	18	308	336	365	17
16. Elbow-fingertip length [54]	448	484	524	23	406	443	483	23
17. Overhead grip reach, sitting [D45]	1221	1310	1401	55	1127	1212	1296	51
18. Overhead grip reach, standing [D42]	1958	2107	2260	92	1808	1947	2094	87

[1]http://cad.fk.um.edu.my/ergonomics/Anthropometry/US.pdf
[2]This example is adapted from Niebel, B. and Freivalds, A. *Methods Standards and Work Design.* New York: McGraw-Hill, 1999.

TABLE 19.1 (continued)

Dimension	Men				Women			
	5th %ile	Mean	95th %ile	SD	5th %ile	Mean	95th %ile	SD
19. Forward grip reach [D21]	693	751	813	37	632	686	744	34
20. Arm length, vertical [D3]	729	790	856	39	662	724	788	38
21. Downward grip reach [D43]	612	666	722	33	557	700	664	33
22. Chest depth [36]	210	243	280	22	209	239	279	21
23. Abdominal depth, sitting [1]	199	236	291	28	185	219	271	26
24. Buttock-knee depth, sitting [26]	569	616	667	30	542	589	640	30
25. Buttock-popliteal depth, sitting [27]	458	500	546	27	440	482	528	27
26. Shoulder breadth (biacromial) [10]	367	397	426	18	333	363	391	17
27. Shoulder breadth (bideltoid) [12]	450	492	535	26	397	433	472	23
28. Hip breadth, sitting [66]	329	367	412	25	343	385	432	27
29. Span [98]	1693	1823	1960	82	1542	1672	1809	81
30. Elbow span	Na	Na	Na	Na	Na	Na	Na	Na
31. Head length [62]	185	197	209	7	176	187	198	6
32. Head breadth [60]	143	152	161	5	137	144	153	5
33. Hand length [59]	179	194	211	10	165	181	197	10
34. Hand breadth [57]	84	90	98	4	73	79	86	4
35. Foot length [51]	249	270	292	13	224	244	265	12
36. Foot breadth [50]	92	101	110	5	82	90	98	5
37. Weight (kg), estimated by Kroemer	58	78	99	13	39	62	85	14

The problem at hand here is to arrange seating in an auditorium such that most individuals will have an unobstructed view of the speaker screen.

Step 1: Determine the body dimensions that are critical in the design (design criteria). Refer to Figure 19.1.

a. Erect sitting height

b. Seated eye heights

Figure 19.1 Design criteria.

Step 2: Define the population being served: U.S. adult males and females.

Step 3: Select a design principle and the percentage of population to be accommodated: Designing for extremes accommodating 95% of the population. The key principle is to allow a 5th percentile female sitting behind a 95th percentile male to have an unimpeded line of sight.

Step 4: Find values from Table 19.1: 5th percentile female seated eye height is 685 mm; 95th percentile male sitting erect is 972 mm. Difference (972 − 685 = 287 mm). For a small female to see over a large male, there must be a rise of 287 mm between two rows. This creates a large slope. Suggestions?

Step 5: Optimization allows for heavy clothes and shoes. Most measurements are made for a naked human.

19.3 PROBLEMS

1. Measure the dimensions of your seat in the classroom. Redesign the seating to fit 95% of females. How would the design affect 50% of males?
2. In a computer lab, suggest a design that will allow most of the users to view the PC and to view their classmates, who are sitting in different rows.
3. What would be the dimensions of a stage podium for your instructor?
4. Because of the *Challenger* disaster, NASA has decided to include a personal escape capability (i.e., a launch compartment) for each space shuttle astronaut. Because space is at a premium, proper anthropometric design is crucial. Also, because of budget restrictions the design is to be nonadjustable (for example, the same design must fit all present and future astronauts, both male and female). For each launch compartment

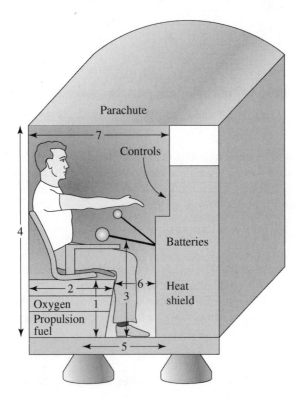

Figure 19.2 Compartment.

feature (Figure 19.2), indicate the body feature used in the design, the design principle used, and the actual value (in inches) to be used in its construction.

a. Height of seat
b. Seat depth
c. Height of joystick
d. Height of compartment
e. Depth of foot area
f. Depth of leg area
g. Depth of chamber
h. Width of compartment
i. Weight limit

Lab 9: Geometric Dimensioning and Tolerancing

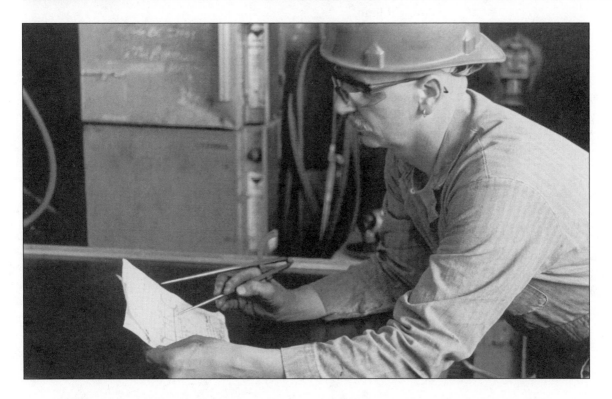

Geometric dimensioning and tolerancing helps the engineer and designer provide clear and concise instructions per tolerance requirements for parts and assemblies as needed.

20.1 PURPOSE

The misperception that CAD models contain all information needed to produce the product has led to the importance of dimensioning and tolerancing. For years the value of geometric dimensioning and tolerancing (GD&T) has been highly regarded due to its ability to provide a clear and concise method of communicating tolerance requirements for parts and assemblies when needed.[1] The purpose of this lab is to introduce students to geometric dimensioning and tolerancing. The information is based on the ASME Y14.5M-1994 standard. The basis of this lab is obtained from The Effective Training Inc. demo.

20.2 THEORY

One of the most important features of design work is the selection of manufacturing tolerance so that a designed part fulfills its function and yet can be fabricated with a minimum of effort. It is important to realize the capabilities of the machines that will be used in the manufacturing and the time and effort to maintain a tolerance. It must be emphasized that in order for the GD&T to be effectively used, the functional requirements of the product must be well formulated. Vague requirements may result in a product that will not function as intended. Similarly, if tolerances are not well detailed, this may also cause a part not to function as intended or make it too expensive. Figure 20.1 shows two samples of outputs for a cup in which (a) the dimensioning and tolerancing standards are not used, and (b) detailed tolerance requirements are stated. It is clear that the output in (a) results in a nonfunctional part while in (b) it results in a functional part. The translations of the tolerancing symbols in (b) are as follows:

1. The top must be in line with the bottom within 2 mm.
2. The cup must not rock; the bottom must be flat within 0.5 mm.
3. The cup must be transparent.
4. The cup must have a smooth surface within -0.4 μm.
5. The volume is 0.5 to 0.6 L.
6. The side angle is approximately 20 degrees ± 1 mm uniform zone.

Designers may have the tendency to incorporate specifications that are more rigid than necessary when developing a product. Figure 20.2 on page 252 shows the relation between the relative cost and the tolerance. As the tolerance requirement becomes smaller, the cost becomes higher. Because of the high cost of tight tolerance, designers need to look closely at their design and adjust functions or components such that the tolerance will not add cost to the product. For example, if a part such as a bearing calls for a tight tolerance (0.005) when installed inside a

[1]Murphy, M. presentation at the NSF workshop at Central Michigan University, 1999.

(a)

(b)

Figure 20.1 Comparison between lack of proper tolerancing and proper tolerancing.

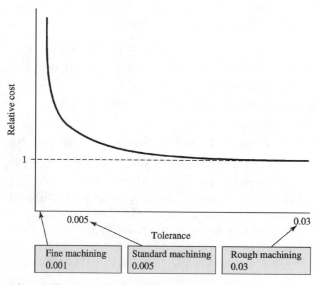

Figure 20.2 Tolerance versus cost.

casing and may function with (0.03) tolerance when placed outside the casing, designers need to change the design to fit the bearing outside the casing.

20.3   PROCEDURE

This lab introduces you to geometric dimensioning and tolerancing. The GD&T Trainer demo CD-ROM can be obtained from Effective Training, Inc. is based on the ASME Y14.5M code. Follow these instructions:

1. Launch the GT demo. There will be a few slides to explain the system requirements and what the program is all about.
2. Click on the course index.
3. Click on Concepts. Remember, this is just a demo version. About three modules will work in this demo.
4. Click on lesson number 3: Modifiers & Symbols. This module will introduce you to the geometric characteristics and symbols. The text that is marked red is clickable and will give you more information.
5. Proceed in this lesson. On page 10 there is a problem that you need to solve. Type that characteristic table as a Word document. You can hand draw the symbols if you can't use Word to reproduce them.
6. Go through the lesson and answer all the questions. Report all your answers in a Word document.
7. Once done with concepts, go back to the course index and click on Forms. Then choose flatness.
8. Go through this tutorial and report your answers.
9. Once you have finished with flatness, go back and take the quiz located on the page for flatness. Report the quiz questions and answers.
10. Exit the demo and start Qr_demo. From the subject index choose position RFS/MMC and report your findings.

21 Lab 10: Reverse Engineering

The engineer must design products, keeping in mind how those products will be manufactured. This chapter introduces the concepts of designing for manufacturing, function analysis, and designing for assembly. Such reverse engineering techniques will not only help the engineer to produce a better product, but will help save manufacturing costs.

21.1 PURPOSE

This lab introduces concepts such as design for manufacturing, function analysis, and design for assembly. In this lab you will disassemble a drill to its essential components and then assemble it back to its original shape.[1] Before you assume responsibility for the drill, make sure that it is in good working condition. The following types of drills are available (we used three types; one type is enough for this activity): Black & Decker DR210K, Skil Twist 2106, Skil Twist XTRA 2207.

21.2 PROCEDURE

1. Identify the type of drill, manufacturer, model number, and performance specifications.
2. Plug in the drill and run it. Listen to how it sounds and feels as it runs. Your senses are a useful qualitative measurement tool. The drill should sound and feel the same after you take it apart and put it back together again.
3. Before you disassemble the drill, list all of the functional requirements that the drill must satisfy. For example, one function could be "hold a drill bit without slipping."
4. Unplug the drill and disassemble it as far as possible. *Hint: Carefully sketch as you disassemble.* Put all parts in a bin.
5. After listing all functional requirements and disassembling the drill, list the physical parts or assemblies that satisfy each function. In the preceding example the function "hold a drill bit without slipping" would be satisfied by a drill chuck. After disassembling the drill to its basic components, you may find out that you didn't list all functions. Make a table showing the functions and the corresponding part or assembly that satisfies that function.
6. Determine the gear ratio. What types of gears are used? What are some other gearing alternatives that could be used in this assembly? Why are they not used?

---

[1]This lab is based on material presented at an NSF workshop, Central Michigan University, 1999.

7. Determine the type of bearing used. Why do you think these types were chosen?

8. Determine what type of motor is used. Why do you think this type is chosen? Describe how the motor in your drill works in terms of basic physical principles. Use illustrations to help your description.

9. Make a bill of material (part list) of your drill. Identify the material used in each part. Also indicate whether the part is an off-the-shelf item or would need to be specifically manufactured for this product (i.e., custom part). What percentage of the total number of parts can be purchased from a parts vendor? Explain your findings.

10. What is the total number of parts in your drill? Compare this number to other drills being dissected. Are there any parts that can be eliminated or combined into a simpler part?

11. Observe the tolerances in the assemblies. Why do you think these tolerances are needed?

12. Reassemble your drill. Be especially careful of the motor brushes.

13. What features of your drill make it easy to assemble? Is there anything that was particularly difficult to reassemble? How would you improve the design of your drill to eliminate this shortcoming? Are there any other improvements to the design that could improve the assembling of the drill? How long would it take for a factory worker to assemble the dull?

14. Once the drill is reassembled and you have no leftover parts, plug it in and try to run it. If it does not sound or feel like it did before you disassembled the drill, or if a strange smell develops, STOP IMMEDIATLY. Unplug the drill and try to determine what went wrong with your assembly steps.

Safety note: Never plug the cord into an electrical outlet while the drill housing is open.

21.3 PROBLEMS

Videotape the lab. Doing so provides visual reinforcement of the parts presented in the lab.

CHAPTER 22

Lab 11: Use of Pro/MECHANICA®
for Structural Analysis

Pro/MECHANICA® from Parametric Technology Corporation is a solid modeling software designed to help the engineer design and test products using virtual prototyping.

22.1 PURPOSE

This lab will introduce you to the use of Pro/MECHANICA® as an analysis tool for structural components. Pro/MECHANICA is a tool available for use in conjunction with the Pro/ENGINEER® software. Pro/ENGINEER® (Pro/E) is a solid modeling software that can also be used to perform virtual prototyping.

22.2 SUPPORTED BEAM

A 2-D simply supported beam by definition has a pin support at one end and a roller support at the other end. The essential feature of a pin support is that it restrains the beam from translating both vertically and horizontally but does not prevent rotation. As a result, a pin joint is capable of developing a reactive force for both vertical and horizontal components, but there will be no moment reaction. A roller support prevents translation in the vertical direction but not in the horizontal direction. The roller also allows free rotation in the same manner as the standard pin joint.

Point and curve constraints on solid elements can cause singular stresses. The following procedure illustrates techniques for modeling a solid simply supported beam using rigid connections:

- Generate the part using Pro/E. The part is a square beam as shown in Figure 22.1.
- Launch MECHANICA Application>MECHANICA. The default units are in (in-lbm-sec). For now assume the default units.
- Select **Structure** from the MECHANICA menu.
- Create datum planes at the ends of the beam. Select insert **Datum Plane.** From the menu shown in Figure 22.2, select **Parallel>Plane.** Select one of the end faces. Repeat for the other face.
- Add rigid connections on each end surface. Click on **Model>Idealizations> Rigid Connection>Create,** and pick one of the end surfaces. Repeat at the other end of the beam.
- Add constraints to each of the end planes. Click on **New Edge/curve constraints,** and pick one of the end surfaces. Fix all Translations as shown in Figure 22.3. Repeat for the other edge; this time leave the Z-Translation free.
- Apply Loads (see Figure 22.4). Click on **Model>Loads>New.**
- Apply Martial Properties **Model>Material.**
- Check Model for errors **Model>Check Model.**
- Select the analysis type. Click on **Model>Analysis>Static>New>Close.**
- Create **Design Study Model>Design Study.**
- Run the analysis. Check the setting. Make sure you know where the files are going to be saved.
- Check the results. Create a results window. Check for the displacement (Figure 22.5) and stresses (Figure 22.6).

Figure 22.1   Square
beam.

Figure 22.2   Menu
manager.

Figure 22.3   Constraints.

Figure 22.4   Load on beam.

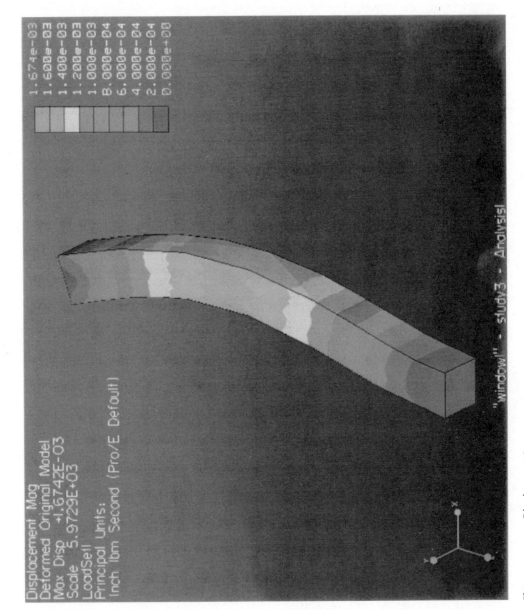

Displacement Mag
Deformed Original Model
Max Disp +1.6742E-03
Scale 5.9729E+03
LoadSet1
Principal Units:
Inch lbm Second (Pro/E Default)

1.674e-03
1.600e-03
1.400e-03
1.200e-03
1.000e-03
8.000e-04
6.000e-04
4.000e-04
2.000e-04
0.000e+00

"window1" - study3 - Analysis1

Figure 22.5   Displacement.

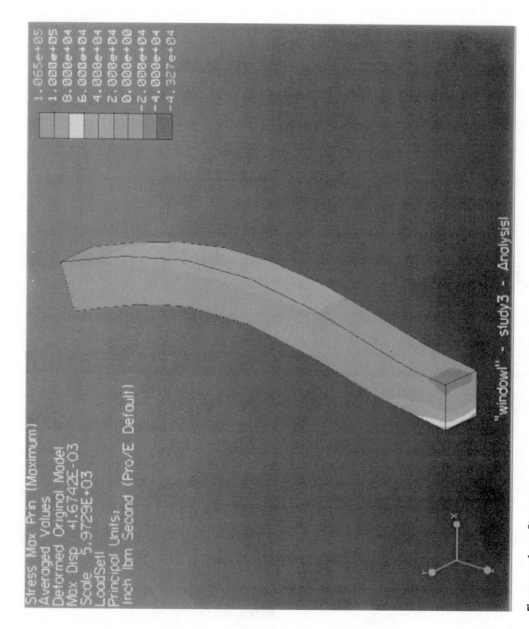

Stress Max Prin (Maximum)
Averaged Values
Deformed Original Model
Max Disp +1.6742E-03
Scale 5.9729E+03
LoadSet1
Principal Units:
Inch lbm Second (Pro/E Default)

1.065e+05
1.000e+05
8.000e+04
6.000e+04
4.000e+04
2.000e+04
0.000e+00
-2.000e+04
-4.000e+04
-4.327e+04

"windowl" - study3 - Analysis1

Figure 22.6 Stresses.

22.3   PROBLEMS

1. Use Pro/MECHANICA to solve for the stress displacement for a shaft of diameter = 2.0 and length = 10.00. A pin on one end and a roller on the other end support the shaft. Solve for the following loads:
   a. Material is steel and the load is uniformly distributed and equal to 10,000.
   b. Material is Al and the load is uniformly distributed and equal to 10,000.
   c. The load is concentrated at the middle and is equal to 100,000 (for both materials).
   d. The load is concentrated at a point 1/3 from the pin joint and is equal to 100,000 (for both materials).
   e. The load is concentrated at a point 2/3 from the pin joint and is equal to 100,000 (for both materials).
   f. Compare the results.
2. Using the beam in the example, change the dimensions of the beam to (10 × 2 × 100). Apply a uniform load of 10,000. (Material is bronze.)
   a. Find the displacement and stresses under this condition.
   b. Assume the analysis is 2-D (use the model type option with plane stress). Find the displacement and stresses; compare with part (a).

Index